MODERN LUMBERJACKING

MODERN LUMBERJACKING

Felling Trees, Using the Right Tools, and Observing Vital Safety Techniques

LEN McDOUGALL

Skyhorse Publishing

TABLE OF CONTENTS

INTRODUCTION

About a month ago, my brother-in-law caused his own death.

It was early in the morning and he was probably as happy as he'd ever been. He and his wife had just moved to a new house in the country, and he was having the time of his life doing a pretty fair imitation of Farmer John. This lifestyle was a longtime fantasy of his, and he was living it to the utmost, even sometimes rising to start a new day at what farmers would call the middle of the night. He had a new tractor, riding mower, and power tools galore with which to exercise his most rural desires.

Among his gasoline-operated paraphernalia was a spanking-new chainsaw that he used to cut down any tree that he could figure out a reason to fell. He, in fact, knew very little about lumberjacking, and had even less experience. But like so many men, he suffered from a gender-based deficit that made him refuse to ask for, or even accept, advice from anyone—after all, to request information would be to admit that he lacked knowledge in the first place.

It was probably not to his benefit in the long run that he'd gotten away with making a truckload of mistakes with the first several trees he felled. He hung a couple of half-fallen trees in the branches of adjacent trees—a classic example of what lumberjacks have always known as a "widow-maker"—and succeeded in bringing them to Earth by simply tying a rope around their trunks and pulling them off their stumps with his tractor.

Those first successes maybe have caused him to believe that all those stories of danger were just invented by tree-cutters who were trying to make themselves look courageous and dashing. This lumberjacking stuff didn't appear to be as hard as it was made out to be! But, sooner or later, the mistakes he made were bound to catch up with him. The one that finally got him manifested itself in the form of a large live oak, about 50 feet tall and weighing as much as three cars, that bordered his driveway.

The sun hadn't fully risen over the hills of his Tennessee home when he fired up his chainsaw and did his best to notch his target tree for that day to fall in the direction he'd desired. Aside from doing a novice-level job on the notch and failing to rope off the tree, his most serious mistake at that point was to regard the job as one that could be started and completed at his leisure. He failed to recognize, or to respect, the powers that he was unleashing in his tools—and in the tree itself.

He proved his lack of understanding by notching the tree early in the morning, then leaving it to sit on its stump while he went into his house for coffee. He came out with his wife an hour later and watched unperturbed as she drove under where the precariously balanced tree was notched to fall. When she returned, an hour later, the tree was down, and her husband was dead beneath it. A neighbor had come across the field that adjoined their houses for a casual visit and found him crushed to the ground with a multi-ton tree across one shoulder.

Forensic authorities figured that the tree had only partially fallen, when its branches caught in the branches of another tree, causing it to hang suspended over the driveway. There is no more classic example of a widowmaker than that. But my brother-in-law clearly didn't know that, or else he would have never considered walking under that giant wooden sword of Damocles to fetch his tractor.

No one knows why he didn't hear the tree suddenly let loose in a gust of wind. Or why, if he did, that he didn't get clear—trees do not fall silently, or with the velocity of a dodge ball. Maybe he didn't recognize the swooshing sound as leaves and branches brushed past one another. Maybe it was a deer-in-the-headlights moment of paralysis.

Whatever the reason, what caused him to remain immobile in the path of a falling tree is inconsequential. The arboreal hammer hit him in the back and bore him inexorably to the ground with the comparative force of a man stepping on a bug, breaking every bone and bursting every organ that was under it. The coroner said that he "never knew what hit him." But that cliché was for the family's sake; death wasn't instantaneous, and he had time to know precisely what had hit him. But when he did, there wasn't a power in heaven or on Earth that could have prevented the outcome.

It's human nature to regard deceased loved ones' lives through rose-colored glasses and to overlook their shortcomings. But when this man's wife told me that what happened was "an act of God," and that "he knew what he was doing," I bit my tongue so hard I tasted blood.

Without a single exception, every logger I've related this story to has shared my supposition that not only did the deceased *not* know what he was doing, but his intelligence is suspect, as he couldn't acknowledge that fact. My brother-in-law's death wasn't an "Act of God." It wasn't even an accident—not in the sense that it was an unavoidable tragedy that no one could have anticipated beforehand. He died from a terminal case of ignorance and a fatal dose of pride.

Which brings us to the purpose of this book. In my own half-century of living in timber country and cutting wood for just about every reason there is to cut wood, I've seen an awful lot of people get hurt awful quickly in an awful lot of different ways from lumberjacking. And, truth be told, I've been one of them a couple of times myself. But the worst injuries didn't happen to me, and I was lucky enough to learn the most important lessons vicariously. It is the focus of this book to take those lessons from my life, from my own experiences and observations, as well as from the experiences and observations of people who are more expert in areas of the timber business than I am, and present them to you in a concise format that will be of real, usable value.

The lessons contained in these pages can be as dark as they are important. There's just no denying that a whole lot of blood was shed, that more or less vital pieces of participants' bodies were lost on occasion, and every so often, someone was killed. All of that's a bit morbid and more than a little depressing. And that doesn't make for a fun-to-read book, does it? So I'm going to make every effort to keep things light. I'm going to salt the lessons with anecdotes that, if I tell them right, should lend a little humor. The lessons related within these pages have all proved themselves to be invaluable in real life many times over and, in a few cases, heeding them might very well keep you, or someone else, out of the hospital. Or out of the morgue. But, regardless of that fact, I very much want you to have fun reading (and learning).

Lumberjacking is adventurous, exciting, and even a little romantic (on a personal note, I've never met a lady who didn't appreciate the embrace of a sweaty axe-slinger). But it's a lie to infer that lumberjacking isn't fraught with danger. According to the Occupational Safety and Health Administration (OSHA), cutting trees is more dangerous than Alaskan crab fishing *and* you can do it at home for a lot less than the cost of a fishing boat. Nothing about lumberjacking isn't inherently dangerous. So be safe, be thoughtful, be cautious, but make sure to have a good time doing it.

A HISTORY OF LUMBERJACKING

A northern Michigan logging camp, circa 1920; every one of these men was tough as a railroad spike, and twice as hard. (Photo courtesy of Wayne Chellis).

It's no exaggeration to say that if there had been no lumberjacks, there would be no civilized world. No lumber, no buildings; no buildings, no blacksmiths, factories, businesses, or houses. No timber, no lumber; no lumberjacks and loggers, no timber. Literally, for want of a lumberjack, cities would have been lost.

But there has never been a dearth of strong, hard young men to fill the ranks when it came to cutting wood. Maybe it's the satisfaction of feeling the sweat on one's brow and the knowledge that, at the end of the day, you deserve whatever pleasures that you might indulge yourself with, and to hell with what anyone else thinks. Compared to lumberjacks, especially those of the nineteenth century and before, roughneck oil drillers are best suited to selling Girl Scout cookies.

The Roughnecks and the Lumberjacks

Perhaps this amusing anecdote will demonstrate the toughness of lumberjacks occurred In 1973, in the little town of Boyne City, Michigan (population 3,000), Shell Oil was pulling a shameful end-run around courts, who had ruled that they could not drill natural gas wells in the protected elk habitat of Pigeon River Valley. Shell drilled anyway, then capped the wells, and paid the pathetically low fines. In the end it was worth it to them. (I got this story from one of their army of defense attorneys, when he was half-drunk.)

The brigade of roughnecks who did the deed for Shell, mostly from states where magnolias grow, judging from their accents, considered themselves to be epitomes of manhood, just about as tough as a human being could get. They were young men who worked hard and liked to raise a bit of hell when their shift ended. They made a practice of descending upon local bars and terrorizing patrons; men trying to re-enact a war that had been settled more than a century earlier. Small town law enforcement tended to cut them maybe too much slack, because the portions of their incomes they spent were prized by businesses in impoverished northern Michigan communities.

The stuff of America. (Photo courtesy Oregon Department of Forestry).

The roughnecks probably should have stayed away from the Tannery Saloon. Named for the old leather tannery that had once occupied that site, this classically redneck bar was a preferred watering hole of loggers. Damage to the establishment was relatively minor, but the oil men never returned.

* * *

Maybe it's the feeling of power that naturally accompanies bringing down a forest goliath with just the muscles of your back and arms. To feel the earth tremble beneath your feet as a dozen tons of timber crashes to the ground, and especially when a tree lands precisely where you said it would. You can't help feeling a sense of power. You grasp just a little of how humans, nature's most anemic and under-endowed species, have been able to transform the world to meet their needs.

Times have changed immensely since the first man hewed through the trunk of a living tree with a knapped stone that he'd tied to a wooden haft with rawhide thongs. The ability to hack down a standing tree was a major milestone for the human race because, in forests where stone and other building materials might have been so limited as to be useless, trees often provided the only suitable building material.

Every culture that had trees learned to make shelters and boats from saplings, bark, and resins, but work with stone axes was limited by the tools (the same holds true today). It required metal-forging, first brass, then iron to fashion axes that could chop down trees large enough to be used as timbers and lumber. Many groups of people didn't reach that stage for themselves, but were bequeathed metal tools that were originally conceived by explorers from elsewhere.

The iron-blade axe was both sword and plowshare. It could clear land for building materials and for crops, but it was a ferocious weapon if a need arose to defend those lands. As a weapon, an axe has no respect for the heaviest armor a strong man can carry. From it was born the more elegant and versatile (and more portable) armor-cleaving broadsword.

Lumberjacking even had a very direct impact on clothing styles. Contrary to the *Braveheart* movie, early Celts, being wood cutters, did not wear kilts. As the tunic-clad Romans noted, and later adopted that style themselves, "barbarian" Celts (and Vikings, and a few other lumberjacking civilizations, which made their sometimes large buildings from timber) wore trousers. Trousers offered protection from cold (original lumberjacks all hailed from frigid climates) and from injuries to legs and knees that goes with such rugged work. Trousers were deemed to be better for riding horseback as well, and the days of the tunic came to an end.

Footwear worn by these primitives was different, too, offering far more coverage than a thong sandal or animal skin bootie. Lumberjacking boots were heavily built to provide armor against rough-barked trunks and jagged, broken branches. Soles were thick, not only to prevent injuries to a wearer's soles, but to facilitate kicking, stomping, and other leg work that aided in bending forest products to human will.

Even gloves originated in the far north, probably as much as protection for lumberjacks as to keep their hands warm. Later, gloves metamorphosed into armored gauntlets, which could safely grasp an opponent's sword blade, and today, into Kevlar™ gloves that can resist a running chainsaw.

Europe and Asia had been exploiting forests for development of their kingdoms and navies long before Jesus was born. So when the superpowers of the day began to civilize the New World at the beginning of the fifteenth century, those first pioneers brought with them what tools they had to conquer the vast forests of America.

By the time Sir Walter Raleigh established the ultimately doomed "lost colony" of Roanoke, in what is now Virginia, in 1587, the ring of Spanish axes had been sounding in other parts of the New World for decades.

When new settlers returned to build the colony of Jamestown, twenty-two years later, they knew what they were in for, literally carving homes from the vast wilderness of America.

Even so, cholera and typhoid nearly wiped out that first successful colony. Jamestown grew and multiplied, spreading up and down the Atlantic seaboard, and the first lumberjacks kept pace with the demand for timber.

As civilized (in hindsight, an arguable term) America grew, so did the need for newspapers and letters from home. In 1690, just after turning pulpwood into paper had become an industry, an entrepreneur named William Rittenhouse began a paper mill near Germantown (now Philadelphia). The timber industry in colonial America was assured.

Early logging activities in Wisconsin. (Photo courtesy of Wisconsin Chamber of Commerce).

As technology improved, and the means to harvest timber with it, more, bigger, and better became bywords of every civilized society.

By the early 1830s, Bangor, Maine, was shipping more commercial timber than any place in the world. In following years, as logging operations moved inland, every waterway that was big enough to float a log was loaded with rafts of unprocessed timber. Horses and oxen hauled cut trees overland on sledges, on unique carts called "big wheels," or just by dragging, if the terrain permitted. Sometimes logs were laid to fashion crude railway systems for transporting other logs.

If there was a river near the cutting site, a "flume" on a downhill slope, or a ditch on level ground, was often dug to connect them—in places like Michigan's Upper Peninsula, these water-filled ditches are features of the landscape today. Riders, armed with "cant hooks" (explained later), would stand atop the logs, clearing jams and snags that might keep the timber from floating smoothly to a mill that was generally built at the mouth of the river.

Being a rider was decidedly dangerous work. Waters were often lethally cold, and riders had to be sure-footed and quick to keep themselves on top of the rafts. Having a log roll over you, and being caught between them, held underwater as the heavy trunks closed together, over your head, has killed a good many young men. That very real hazard spawned the sport of log-rolling, which has become a featured entertainment of modern lumberjacking festivals.

On many rivers, and especially on barges crossing the Great Lakes, logs became waterlogged and sank. In some rivers, currents caused one end to get stuck in the mud, and the logs are still there today, more than a hundred years later, posing hazards to boats. The Tahquamenon River, of Henry Wadsworth Longfellow's *Song of Hiawatha* fame, is littered with such navigational hazards. In some instances, like a company that calls itself Superior Timbers, logs that have been sunk, some since before the Revolutionary War, are being recovered and sold commercially to high-end markets.

As civilization spread west, lumberjacks inevitably followed. When the Homestead Act was passed in 1862, with the promise of 160 acres per family, lumberjacks were in high demand, as much of the deeded acreage was heavily timbered; it needed to be cleared for farming, and lumber hewn from those acres was needed for building.

Harbor Springs, Michigan, is one example of numerous boom towns that were born not from gold or silver, but timber.

Until the 1940s, when manual lumberjacking went into decline, loggers lived a cold, lonely life in remote logging camps that were snowbound in winter and bug-infested in summer. Work days were from sunup to sundown, labors were dangerous and hard, and pleasures were few. Lumberjacks rarely bathed during the winter months, as laundry services were almost nonexistent and medical care was primitive. Infections and disease were common, lice infestation was almost normal. Hard living begets hard men, and soon lumberjacks had acquired a fearsome reputation that remains today—in most instances it was (and is) merited.

Logging is a year-round occupation, even in places the normally get 240 inches of snow.

A logging train, taking jack pine logs to the mill, where they'll become Oriented Strand Board.

As logging camps grew into towns, lumberjacks necessarily became more genteel, marrying and raising families. This brought in schools, churches, and commerce, and many communities grew into large urban centers, like Duluth, Minnesota, or Seattle, Washington.

According to *Forbes* magazine and other official sources (because it takes a statistician to determine what other people already know), being a lumberjack remains the toughest and most dangerous occupation on Earth. More than one in a thousand loggers is killed on the job, a mortality rate about thirty times higher than most industries.

The goal of this book is to reduce that mortality rate to zero. If you don't have to do it for your living, lumberjacking can actually be fun. So be safe, have fun, and let the (wood) chips fall where they may.

Turn-of-the-century lumberjacks taking a lunch, or "nosebucket" break. (Photo courtesy of Wayne Chellis).

Early loggers sometimes overloaded sleds and wagons as a test of their strength - and for publicity shots, like this one.. (Photo courtesy of Wayne Chellis).

Bill Chellis, circa 1920; it was said that this lumberjack never went anywhere without his whiskey jug. (Photo courtesy of Wayne Chellis).

Like gold-rush towns, many a logging camp grew from a single stump to become a thriving community. (Photo courtesy of Wayne Chellis).

Off-duty lumberjacks, bringing home their version of take-out. Note that the author has known a few hunters who had missing fingers from posing like this with a gun. (Photo courtesy of Wayne Chellis).

Taking a break from the brutally hard labor of pulling stumps (and mugging for the camera). (Photo courtesy of Wayne Chellis).

Living in a logging camp at the start of the twentieth century was a hard life, best suited for strong, young men.
(Photo courtesy of Wayne Chellis).

Lumber mill, where timber became lumber. (Photo courtesy of Wayne Chellis).

Doing lunch at lumber camp, circa 1920. (Photo courtesy of Wayne Chellis).

Lumberjacks at a logging camp near Harbor Springs, Michigan, circa 1920. (Photo courtesy of Wayne Chellis).

A Big Wheel of the type that gave Dead Man's Hill, in Michigan's Jordan River Valley, its name after lumberjack Stanley

(Big Swede) Graczyck was killed by one in 1910.

Lumberjacks at logging camp, circa 1920.

Even lumberjacks at he turn of the century took time for a little recreational sports.

Women, men, and children—tough people all—building the community, which would later became a small city.

TOOLS OF LUMBERJACKING

Like every hands-on profession, every occupation, the expertise of a lumberjack is directly determined by his tools. Raw talent, physical aptitude, and personal toughness, even hard-won skill can carry a wood-cutter only so far. A lumberjack, as the old saying goes, is only as good as his tools.

That means that the lumberjack of today can be very good, indeed. From laser cut pruning saws and hammer-forged stainless-steel axes to cordless chainsaws and collapsible ladders, the tools available to modern wood cutters are the stuff of science fiction for generations past.

Chainsaws

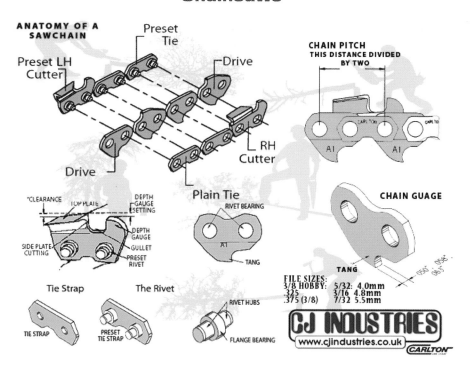

Anatomy of a saw chain (courtesy of CJ Industries).

This is the tool that turned lumberjacks of old into today's loggers, and brought the tree-harvesting occupation into the modern era. Within just a few years of its creation, the *thunk* of an axe biting into tree trunks, and the *buzz-buzz-buzz* of a man-powered crosscut saw had been replaced by the mechanical drone of a gasoline-powered piston engine.

A lumber camp-turned-museum, commemorating the place that became the town of Newberry, Michigan, in Michigan's wild Upper Peninsula..

For help in composing this chapter, I turned to Pete at PM Small Engine, in Newberry, Michigan, located in the very heart of timber country. Pete is a dyed-in-the-wool Yooper who doesn't feel comfortable in the limelight (even though he's enough of a businessman to want his store publicized). Pete is also a magician when it comes to making a chainsaw sing, and he's no slouch when it comes to using one, either.

When it comes to chainsaws, Pete is hesitant to recommend a particular brand, because many logos are, in fact, owned by the same few conglomerates. He refers to many of the brands found in department stores these days as "dimestore chainsaws." Even among the major brands, whose commercial-grade saws are popular with professional loggers, he expressed disappointment at seeing "things break that shouldn't be breaking."

Pete's place; the author's chainsaw expert's store; adorned with antique log tongs, sawmill blades, and an old anvil. Like a typical Yooper lumberjack, Pete shuns the limelight, but the author wishes to thank him for contributing his expertise to this book.

The Chainsaw's History

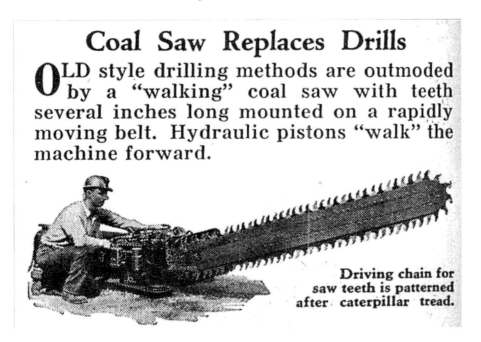

Coal Saw Replaces Drills

OLD style drilling methods are outmoded by a "walking" coal saw with teeth several inches long mounted on a rapidly moving belt. Hydraulic pistons "walk" the machine forward.

Driving chain for saw teeth is patterned after caterpillar tread.

An example of the utility in the chainsaw concept.

The first machine to employ a rotating chain to cut was developed in the late eighteenth century by two Scottish doctors, John Aitken and James Jeffray. They didn't work together on the invention, but seem to have both developed the concept simultaneously.

The chainsaw's original purpose was medical. It was used to excise necrotic (dead) bone and to perform autopsies, and for a time was also used to perform *symphysiotomies*. This latter procedure was once a component of caesarean births, used to break and separate a mother's pelvic bone to widen the birth canal; this controversial practice was discontinued in 1984, and today is considered as archaic as sawing off wounded limbs.

The first chain hand saw debuted in 1783, and employed a fine serrated link chain. It was featured in Aitken's *Principles of Midwifery, or Puerperal Medicine* (1785) and used by him in his dissecting room on cadavers.

Jeffray laid claim to the invention about the same time, though his version didn't see use until 1790 before he was able to have it produced. In his co-authored paper, *Cases of the Excision of Carious Joints*, Jeffray explained that the chain saw permitted a smaller incision, and protected the adjoining neurovascular bundle. Mechanized versions followed, until the latter nineteenth century, when it was succeeded by the Gigli Twisted Wire Saw.

Innovations:

1785: A hand-driven chainsaw, with a fine serrated chain, is used to excise diseased bone; it appears in John Aitken's *Principles of Midwifery, or Puerperal Medicine*.

1926: Andreas Stihl patents two 2-man chainsaws—a 116-pound electric-motor version, and later, in 1929, a 139-pound gasoline-engine model. By 1937, the weight had been trimmed to a mere 88 pounds. These are imported as spoils of war by the United States Army in 1941.

1945: Chain saws are still too heavy to be practical in the deep woods; saws are sometimes wheeled and require two operators. Aluminum alloys and stamped-steel components (wartime innovations), reduce size and weight to practical, man-portable levels.

1947: Inspired by the 2-way chewing action of a Timber Beetle larva, Joseph Buford Cox invents the Cox Chipper Chain.

1949: McCulloch Motors Corporation unveils their 25-pound Model 3-25 as the "World's Lightest Chain Saw."

Chainsaws: The Controversy

Maybe it's because *Forbes* magazine and the Occupational Safety and Health Administration (OSHA) have declared lumberjacking and logging to be the most dangerous occupations in the world, but most manufacturers of wood-cutting tools and equipment are jumpy as cats in a dog kennel. Aside from paid advertisements, the contents of which they control exclusively, most manufacturers of timber-management tools—even axes—are inordinately skittish of writers who aren't directly under their control.

One can only presume that this apparently self-defeating attitude (what manufacturer or retailer *doesn't* want to advertise its products with independent—i.e., more credible—professional reviewers?) is because a surprising percentage of manufacturers appear to be very afraid that, in today's litigation-happy culture, they'd face lawsuits over even self-inflicted injuries.

Unfortunately, I've seen that such things have happened numerous times in the past. As far back as the early 1980s, many a would-be lumberjack, drafted and put into gainful employment through Ronald Reagan's Comprehensive Employment and Training Act (CETA), suffered injury through his own actions, sued for damages, and actually won a cash settlement. Today, in a world where firearms manufacturers are blamed by many for misuse of their products by others, there's no reason to expect a sudden epidemic of logic or responsibility to ensue.

One of the two top chainsaw manufacturers made its participation in this book contingent on its lawyers approving everything that I wrote about their products. That wasn't ever going to happen, and only a lawyer would think that any journalist worthy of the title would agree to such a prostitution of integrity. Fortunately, plenty of manufacturers were still eager to participate so the effect on the quality of this book is insignificant, but attitudes are surprising.

Politics notwithstanding, choosing between chainsaw makes is very much like asking a roomful of people which brand of automobile they prefer. All chainsaws are designed to cut wood, and all of them perform as advertised. For that reason, this book refrains from recommending any specific chainsaw, except where some critical point of excellence makes it obligatory to the reader that I do so. Comprehensiveness demands reviewing the performance of as many makes, models, and tool types as was possible, and in a few instances—like recently introduced battery-operated cordless models—the the requirement for completeness turned out be a pretty cool blessing.

More important than a manufacturer's decal is that you get the proper chainsaw for what you intend to do, and that you use it properly.

Types of Chainsaws

Tended to and maintained like the valuable tool that it is, a chainsaw can serve its owner well enough to become an heirloom.

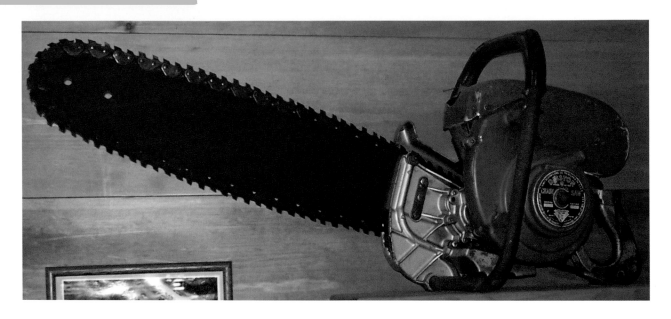

One of the first man-portable chainsaws; note that it has a gearbox, and that its chain is comprised of all cutting teeth, with no rakers or depth gauges..

Another of the first forester saws, this one not truly a chainsaw, because it cuts with a more conventional reciprocating saw blade.

Begin with selecting the proper *size* of chainsaw. This is the most important step in determining which saw fits your needs, because, regardless of manufacturer, some will just not do the job you require of them. However, more critically and more commonly, some of them are far more powerful than an average homeowner needs—like choosing a motorcycle, bigger is not always better, and sometimes bigger is worse.

When motorcycles became almost a fad at the turn of the millennium, there was a dangerous trend toward buying bikes that were too big and heavy for their often novice riders to handle—or even to lift back onto their wheels when they fell over. Too big a bike is clumsy and dangerous, but too big a saw can turn into a real hardship.

Gasoline

More than one chainsaw manufacturer claims to have invented the gasoline-burning chainsaw—most of them around 1920. What is not in dispute is that in 1926, a thirty-year-old German engineer named Andreas Stihl patented the first gasoline-powered chainsaw, a huge 2-man contraption that weighed more than 100 pounds, which he aptly termed a "tree-felling machine."

What followed was a steady evolution of innovation and improvement, and now gas-powered chainsaws have become lightweight, powerful, and efficient enough to be indispensible field tools.

My expert, Pete, says that any gas-powered chainsaw with an engine size of less than 40cc is suitable only for pruning, or for work on a Christmas tree farm. (Chainsaws in the 20cc–30cc range are often referred to as "pocket" chainsaws by professionals.) Many people are attracted to a smaller sized saw because smaller appears to be safer. This is a misconception that probably causes more injuries than it prevents, as the perception of safety is an illusion—like parents who used to (this isn't such a common thing these days) buy little Johnny a .22 because it was a "safe" gun. A small saw is still potentially very, very dangerous, and a user should never let his or her guard down for a moment.

At the same time, a smaller gasoline-powered saw *is* easier to control, with a less violent kickback. A saw in the 20cc–30cc range, with a 10- to 14-inch bar, is usually ideal for suburbanites who need to buck an armful of fireplace fuel or a few storm-downed tree limbs. Every chainsaw is capable of chewing off a limb, but most folks don't require the power to section a telephone pole.

The best-selling rural and semi-professional chainsaws have engines of 40cc to 50cc, and cutting bar lengths of 16 to 20 inches. This is the preferred size for cutting a winter's supply of firewood, for felling trees, and for general all-around use. This is a powerful saw, capable of strong "kickback"—a phenomenon that sometimes occurs when the bar is pinched, when the chain makes sudden contact with a hard knot, while cutting through multiple pieces of wood (very hazardous), when a chain is unevenly sharpened, or when cutting with the tip of the bar, which is never recommended.

Professional-level chainsaws, built for heavy, prolonged use, such as felling large stands of lumber-grade timber, typically have engines from about 60cc to more than 100cc, with cutting bars more than 22 inches long. These are very big saws with extremely powerful engines, and are not suggested for anyone who doesn't cut timber for a living.

Electric

Urban dwellers whose needs are limited to trimming an occasional limb might consider an electric chainsaw, with a 10- to 14-inch bar. Corded electric-motor chainsaws are of limited use because they need to plugged into a 110-volt AC outlet, and that, of course, limits their traveling distance to the length of an extension cord.

But AC chainsaws are far from useless; for cutting fireplace wood, split-rail fence posts, trimming dead branches from trees, cleaning up wind damage to trees and shrubs—even some remodeling jobs—a corded chainsaw is just another useful power tool.

On some farms, where owners butcher their own cattle, a small cord chainsaw is sometimes used to split and quarter beeves. A "side" of beef is created by splitting a carcasses' spine lengthwise, from neck to tail; a daunting task that is most often performed with a reciprocating saw, but a small electric chainsaw isn't unusual. A gasoline-powered chainsaw would contaminate the meat with petrochemicals.

An electric chainsaw that's used for cutting meat is usually never used for any other purpose, and its chain-oiler must always be lubricated using vegetable-based cutting oil, like **Accu-Lube**® food-grade lubricant. Clean the saw thoroughly after every use with soap, water, and chlorine bleach. Take care not to get water in the motor, and make certain the tool is completely dry, inside and out, before storing.

When using a corded chainsaw, be especially mindful of its power cord. It's very easy to cut this life-link in half with just a touch of the chain. If you do, loss of power might be the least worry; the plastic body of most models protects a user from electrical shock, but a short-circuit with the metal chain bar could damage the motor, perhaps even burning out an armature winding, and effectively destroying the motor. Not to mention the danger of short-circuiting yourself.

Battery

The ideal chainsaw is battery-operated: quiet, pollution-free, powerful, yet environmentally friendly—in today's terminology, "green." I asked my chainsaw expert, Pete, what he thought of battery-operated chainsaws; he screwed up his face in that trademark grimace that he displays when something doesn't sit right with him, and said, "They're toys."

But no expert can know everything. A lot of horse experts predicted that the earliest motor carriages were just a fad that would die out when people realized that equestrian modes of transportation offered many real advantages over the rattling, choking, unreliable, over-heating automobile. People are uncomfortable with change; new is different, and different is unfamiliar. Unfamiliar sometimes leads to errors, and mistakes are sometimes dangerous to us.

That same resistance to change has been played out in every succeeding generation throughout human history, and naysayers—of which I have been known to be one—are usually, inevitably wrong. They fail to take into account the creativity of human minds, and our innate drive to improve everything. The first submarine was a disaster, and the Wright brothers' first airplane was virtually un-flyable, but those machines today have evolved beyond the wildest imaginations of their creators. It's a trademark of humanity that every invention, from can-openers to coat-hangers, is subject to continual improvements.

Although not yet capable of replacing gasoline-powered chainsaws entirely, cordless electric models have reached the point where they'll satisfy the needs of all but full-time loggers.

In 2011, Stihl debuted its MSA 160 C-BQ battery-powered saw. Its lithium-ion battery provided up to thirty-five minutes of cutting time. It was, indeed, a novelty. But it was also a start. Expensive, too limited to be of real use, even in a suburban environment, that first Stihl cordless chainsaw was simply infeasible for even modest lumberjacking work.

For evaluation in this book, we selected two battery-operated chainsaws, which we judged (correctly, as it turned out) to be as good as or better than the rest.

The first was a Ryobi 14-inch model with a brushless motor powered by a 40-volt lithium-ion rechargeable battery pack. The saw had originally hit the consumer market with a three-year warranty, but an insert included with package literature informed us that the warranty had been increased to five years, which spoke well for the unit's performance in the field. The saw alone retails for about $200, with battery and charger about $300.

The second was an Echo 16-inch "Professional" model with a brushless motor, driven by a 58-volt lithium-ion battery pack. Clearly heavier and larger than the Ryobi, the Echo boasts a five-year warranty for personal use, two years if used professionally. That latter warranty speaks volumes for the heavy-duty design of this chainsaw. Retail for saw, battery pack, and charger is approximately $600.

A cordless chainsaw—no ear protection needed, and it bucked more than enough 15-inch logs to fill a three-quarter ton pickup, without running out of power.

This cordless 40V Ryobi chainsaw proved its worth in the vast forests of Michigan's Upper Peninsula.

The first user of the Ryobi probably said it best, after using it to buck a dozen 10-inch logs: "I'm impressed."

In fact, cordless chainsaws still do not compare directly with gasoline-powered models. Yet. Battery life depends on the amount of work being performed, ambient temperatures (cold weather seriously reduces a battery's life expectancy) and other factors—like whether timber is hardwood or softwood, green or seasoned . . . but the Echo easily bucked a dozen 3-foot lengths of green pine whose diameter pushed the 16-inch bar to its limit, and still had more than enough power remaining to cut a dozen lengths of 8- to 10-inch wood. In summary, it cut about half a cord of wood before its battery was depleted. The Ryobi, slightly less.

There is no "fade" with lithium-ion batteries, and when a power cell is depleted, the saw just stops running; the motor doesn't slow down or drag, it just ceases to function. Recharge time—a push-button fuel meter is built right into the batteries—is just one hour.

If you're a professional lumberjack—or if you're setting out to the woods to cut a winter's supply of firewood—a battery-powered chainsaw isn't yet evolved enough to suit your needs. Yet.

But environmentally friendly saws seem to be the wave of the future. If you're not going to cut many cords of firewood, if you want a clean, non-flammable saw that stores in a basement until it's needed to cut a few sticks of wood, or clear away storm damage, or you want a saw that can be carried conveniently in your car's trunk, then the right cordless chainsaw might be the ideal.

Hand Chainsaw

Probably one of the coolest inventions for survival-types in the latter twentieth century was the muscle-powered chain saw. This handy tool rolls up to fit into a pocket, yet it gives you an ability to cleanly cut through a tree trunk that's more than a foot in diameter.

A hand-drawn chainsaw.

Hand-drawn chainsaws are pocket-size (usually with their own carrying pouch), and useful enough to be must-have tools for every DIY lumberjack's kit.

Unlike a motorized chainsaw, which cuts in one direction only, the manual chainsaw employs a specialized cutting chain that has teeth facing in both directions, so that it can be sawed back and forth, cutting in both directions.

This chainsaw is not a first choice for taking down standing timber or for cutting cord wood, but if you have a dead branch that needs to be removed from a tree in your yard this is a convenient tool to have in a tool box or glove box. It's also a pretty handy saw for hunters who need to clear shooting lanes, or campers who need firewood.

Specialty Chainsaws

In the early '70s, when everyone I knew was a good ol' boy and a full-time or part-time lumberjack, a fellow I know, who'd just experienced a financial windfall, showed up, showing off his brand-new sedan.

Problem was, he'd wanted a convertible, but there hadn't been one available when he'd decided to buy the car with money that was almost literally burning its way through his pocket.

Alcohol has a way of making problems look surmountable, and the more the new car owner imbibed with his old friends, the more ideas they came up with for remedying this one. Somebody suggested that the roof could be removed using a chainsaw. Someone else produced a chainsaw, and before you could finish a beer, that hardtop automobile was roofless.

In the cold light of morning, it didn't seem like the customizing effort had been such a good idea.

Now, that sort of use isn't suggested, but it does demonstrate that a rotating chain is an efficient cutting design.

Pole Saws

If the work is light enough to permit it or if you need to cut overhead (DO NOT CUT ABOVE YOUR HEAD WITH A CHAINSAW—that cannot be overstressed), you might opt for a pole pruner. Available in hand-saw types, too, this tool is generally an undersized chain saw with an 8-inch bar, affixed to a pole that enables it to overhead branches. Pole saws are small, but generally fitted with an engine of about 30cc or less.

Pole saws are powerful for their size and can be very dangerous if misused. Accord them the respect that you'd give a full-size chainsaw, including wearing chainsaw-resistant gloves and clothing.

Anecdote: A Lesson About Why You Never Cut over Your Head with a Chainsaw

The young man from Boyne City, Michigan—the one-time lumberjacking town where I spent my youth—knew better. He'd been running a chainsaw since he'd been a teenager, and that meant that he had at least a decade of experience, necessarily cutting tons of firewood every autumn, that should have told him to be more careful.

But we were deep in the woods, off a vague logging road that was marked only by twin ruts pressed deeply into the loam—visible only a few months of the year when ferns and grass didn't hide its existence.

Like this autumn afternoon, when we'd followed it to a hidden stand of American Elm. The trees have all but disappeared from Michigan's landscape today, due a secondary fungus caused by an invasive bark-beetle infestation—and spread unwittingly by those of us who looked on the large standing dead trees as sources of firewood. There was no bottom to the environmental well in those days, and we were all incredibly short-sighted.

The day didn't begin remarkably different than any other day: three friends, in a neighborly camaraderie that's not uncommon among people who live simply and do without to get along, had gathered together to help cut one another's winter firewood supply. There was beer, companionship, and someone had a quarter-ounce of weed to help make the day go by faster.

It's a feature of trees that they grow in response to their unique environments, which is to say, unpredictably. Snow, especially, twists and bends young trees every which way. Sometimes branches can extend in all different directions, sometimes at once, and very often, they're in the way.

So it was with the thick, dead overhead branch that was blocking the young man from making his intended cut on the trunk. Per standard procedure, he placed his chainsaw at the junction—the crotch—where limb intersected trunk, and began cutting with the tip of his saw.

Kickback is a probability when you cut with the tip of a chainsaw, but a lot of experienced sawyers do it, anyway. In fact, chainsaw artists, who carve some pretty magnificent sculptures, using a variety of modified chainsaws, work with the tips of their chainsaws almost exclusively. Like handling a dirt bike over rough terrain or rolling with the recoil of a firearm on full-automatic, they've learned to work with kickback.

But this young man was hardly a chainsaw artist, and he knew better. However, like young men everywhere, he suffered from an illusion of indestructibility. Halfway through the branch, the saw kicked back and, in his off-balance overhead position, he was unable to control it. The cutting chain, whirring at full revolutions (because his ungainly orientation caused the kickback to exert leverage on his trigger finger) shredded through his jacket to the flesh beneath. Before it forced him to release the trigger, the saw's bar was through his collar bone and deep into a shoulder blade.

For such wounds, the advice from paramedics is to leave a foreign object in place. But we didn't know such things in the '70s and, in any case, it would have been impossible to leave that 10-pound cutting tool hanging there, embedded deep into his shoulder, while we bounced out of the woods. An ambulance couldn't reach us, even if we'd been able to call one, and it was half a mile to the nearest asphalt road.

He made the trip to the hospital, 30 miles distant, with a greasy rag clamped over the wound, held there by the iron grip of a lumberjack. He survived, but not without substantial blood loss, immediate surgery, permanent tendon and bone damage, and a considerable period of healing.

This particular anecdote is presented here to demonstrate why you should never cut over your head with a chainsaw—that job is why pole saws were invented—but the lessons it teaches to any-one who'll listen are numerous. Kickbacks are more likely when cutting with the tip of a saw, always keep a firm, solid grip on your saw, always carry a modern first-aid kit, with Quik-Clot™-type clotting sponges, always take advantage of the life-saving capability of having a cell phone . . . and never forget that a tool made to grind through wood will find little resistance from body parts.

Concrete Saws

Fitted with specialty chains—like the comparatively blunt diamond-abrasive chains that are used primarily for cutting apart hardened concrete—the concept of chain-sawing shows itself to be ingenious.

Concrete saws are generally a bit more powerful than wood-cutting saws, but aside from their specialty chains, they aren't noticeably different designs than their tree-sawing counterparts.

CHAINSAW ANATOMY AND MAINTENANCE

Chainsaw Anatomy

Throttle

The *throttle* regulates a saw's RPMs by increasing or decreasing the volume of fuel to the cylinders. The chainsaw will stop the chain when pressure on the throttle is released. Its design almost always resembles a trigger, which must be fully depressed to bring the chain up to cutting speed.

A chainsaw is intended to work at maximum RPMs. It has two settings: idle, with the trigger completely released, and full-speed, with the trigger fully depressed. Do not try to run your saw at any speed in between.

A cold saw may refuse to idle, and sawyers often release and depress the trigger repeatedly, pumping it until the spark plug is hot enough to keep it running. If that problem persists, a new spark plug may be in order. If replacing the plug doesn't do it, the throttle screw on the carburetor may need to be adjusted: a job best left to a technician.

Throttle Interlock

The *throttle interlock* locking mechanism prevents the throttle from activating until the interlock is depressed. This safety device is a lever, located on the handle, opposite the throttle trigger. It must be depressed by a user's palm to release the throttle, and enable it to be depressed, forcing an operator to take a firm grip on the operating handle before the chain can be engaged.

Chain Brake

A *chain brake* became an OSHA-required safety feature in February 1995. A chain brake stops movement of a chain when the throttle trigger is released—as when the saw kicks back. It essentially causes the chain to drag when the throttle trigger is released, preventing it from free-wheeling and possibly cutting skin or grabbing clothing.

Personally, I have mixed feelings about this feature, because it tends to cause novices to believe that they can safely touch the chain to a portion of their anatomy, so long as they don't apply power with the throttle trigger. This belief is fallacious: NEVER touch the chain, or let it come near any part of your body. Tell yourself that the chain and bar are red-hot, and that they'll burn you badly if they make contact with your skin.

At the same time, my expert, Pete, tells me that he believes a chain brake to be one of the best innovations to lumberjacking since Kevlar™ chaps. That's good enough for me.

Chain Catcher

The *chain catcher* is a metal or plastic guard located on a saw's body at its bottom, where the underside of a chain enters the body to rotate around its drive sprocket. This is another safety feature that's designed to catch and wad-up a broken or derailed chain, preventing it from striking the operator.

The best use of a chain catcher is to not use it at all. Keep your chain adjusted just loose enough that you can pull it out to the bottom of one drive tooth—not so snug that it can't rotate freely (with the saw turned off, not running) by hand, but tight enough that it cannot jump out of the bar's guide groove.

Again, do not rely on a chain catcher. It's still possible for a broken or derailed chain to whip against an operator's hand, sometimes lacerating it severely. For this reason, leather or otherwise heavy, work-grade gloves should be worn religiously when running a chainsaw.

Hand Guard

The *hand guard* is a defensive shield located atop a saw's body, just before the bar and chain. This guard also prevents injury from a derailed chain, but probably the most important protection comes from keeping an operator's hand from going forward to contact a chain. This includes helping to guard against kickback.

Again—and this advice cannot be overstressed—the best use of any guard or safety feature is to not need it at all. Always keep a firm grip on your saw. Wear gloves, helmet, chaps—any protection you can get. And stop when you become fatigued, because fatigue generates mistakes.

Muffler

The *muffler* is a hearing-protection device used on chainsaws to reduce engine noise. Essentially an insulated gas-expansion chamber, a muffler prevents a gasoline-burning piston engine from being so loud that it would quickly damage a chainsaw operator's inner ear.

A chainsaw with a holed or missing muffler shouldn't be considered in operating condition. Super-heated exhaust gases coming directly from the engine are dangerous, capable of burning skin or even setting fires, and the noise level is hazardous.

Even with a good muffler, hearing protection is recommended when operating a chainsaw. This is an often overlooked safety requirement, but some lumberjacking helmets, like 3M's pretty awesome PELTOR™ Lumberjacking Kit, incorporates hard hat, ear muffs, and visor, all in one convenient unit.

Anti-Vibration Handle

A shock- and vibration-absorbing *anti-vibration handle system* is recommended by OSHA to limit ergonomic stress to an operator's hands, arms, elbow, and shoulder joints.

For an average homeowner-lumberjack, the arthritis and other joint-related maladies that may be caused by the constant juddering of a hand-held piston engine on a daily basis aren't a major concern. But every logger can tell you running a chainsaw is tiring, and getting tired leads to mistakes in judgment. For that reason alone, an anti-vibration handle is a must-have.

Decompression Valve

A *decompression valve*, located above the piston cylinder on mostly newer saws, isn't really a safety feature, but releases compression inside the chamber, making it easier to pull the starting cord.

This feature, of course, means that the yanking needed to start an engine is less violent, allowing more control of the saw, with less chance for a slip. And it's less tiring. Even a small reduction in energy expended while cutting wood is advantageous, as it decreases the possibility of a fatigue-induced mistake.

Clutch

The *clutch*, attached to the chain's driving sprocket, at the side of an engine, is the friction-operated connector that controls the transfer of power from crankshaft to chain. A pinched bar, with bar and chain trapped between two halves of a large log, is a regular problem, and when that happens, a clutch's job is to stop the chain.

A clutch's purpose is twofold: Firstly, it limits the amount of force that can be transferred from an engine to the chain it drives, thereby protecting a powerful engine from damaging itself. Secondly, by stopping the chain when it makes contact with an essentially immovable object (large steel spikes, driven into trees as anchors, and for other purposes, are not uncommon), a clutch protects the operator from injuries that might result from a broken chain or flying parts. Clutches *can* wear out, however. If your saw appears to "slip" when you apply it to a piece of wood, or the chain stops rotating before it cuts, it's clutch might be defective or worn out. To prolong the life of your saw's clutch, <u>release the throttle trigger immediately when the chain stops turning</u>. If the problem persists, take your chainsaw to a qualified technician.

Extracating a Pinched Chainsaw Bar

If you operate a chainsaw, you're going to get its bar pinched between the wood you're trying to cut. That dilemma is as much a part of the job as keeping oil in a saw's chain oiler, and it's going to happen to you in an almost uncountable number of ways.

Getting a bar pinched isn't serious in itself, but getting it free requires a bit of finesse or it's very possible to completely ruin both the bar and chain—and even to wreck the chainsaw entirely.

If you rev-up your chainsaw and its chain refuses to move—the engine just gives out with a low, trombone-like moan—shut down the machine immediately. That sound means that the engine is slipping against its clutch—as it's supposed to do—and further revving the motor is just causing unnecessary wear on clutch plates.

First, with engine shut off, try pulling straight backward with the saw. This will almost certainly not work; the bar will slide to the end of its nose where the wider cutting chain, whose teeth are engineered to bite into wood, will prohibit pulling it free.

If a log is cut through sufficiently when it pinches the bar, you might be able to widen the cut by pushing against the end of the log or by levering the cut itself with a pry bar until the cut opens wide enough to let the chain slip free. Never pry against the bar itself and never exert any outside force of any kind, in any direction, against a cutting chain.

If gently pulling and prying isn't adequate to free a trapped bar, it's time to operate with an axe or, if you feel less than comfortable with a 3.5+ pound head (because, make no bones about it, this is a delicate operation that demands skill), use a smaller hatchet. The objective, at this point, is to remove wood around the stuck bar until there's no longer anything pinning it.

The procedure is the same as for chopping a log in two; that is, to remove wood one chip at a time until you've created a notch that leaves bar and chain free. Except, in this case, great pains must be taken to avoid striking any part of the chainsaw.

First, "set" your cut by driving in the axe's blade at an inward angle on either side of the pinched bar. Space between the setting cuts should be equal, minimally, to the depth at which the chainsaw's bar is buried. The slightly exaggerated distance between setting cuts helps to ensure that your axe won't strike the embedded bar; it might seem that you're trying to remove extraordinarily large chips, but remember that there's no support for the grain on the side where the bar is stuck, because that side is, in fact, cut through.

Strike the first blow on either side at an inward angle, roughly 45 to 60 degrees. As when chopping a log or tree trunk, your weaker hand is your guiding hand and grips the haft just above its toe at the end. Your stronger hand begins high up near the axe's head.

Begin your swing by placing the axe over your strong-side shoulder—as if it were a baseball bat—then drive it downward, allowing your topmost driving hand to slide smoothly toward the end of the haft until it's positioned atop the opposite hand at the toe of the handle; doing this generates maximum power for the least expenditure of energy on your part.

Note that this is also the chopping procedure used for halving a log or felling a standing tree except that, in this case, you strike only at the outsides of the setting cut to avoid hitting the stuck chainsaw. Twist the axe handle after each strike and you'll be rewarded with large chips of wood coming away from, and freeing, the embedded tool.

Sooner or later (probably sooner), your chainsaw will become pinched in a vise-like grip by the very wood it had been cutting. How you address that dilemma is one of defining traits of a woodsman.

"Setting" the cuts that will free a stuck chainsaw.

Removing chips that will, when enough have been taken out, will free a pinched chainsaw.

Although care must be taken to avoid striking a pinched chainsaw bar, freeing it with an axe is actually a very simple process.

Removing chips from the opposite side of a pinched bar, using an axe.

Avoid striking the saw's bar, and chips will chop free easily and quickly.

Cutting large wood is simply a precursor to getting your chainsaw pinched; you need to know how to free your saw when this happens.

Flywheel

The *flywheel* is a weighted wheel that helps to control engine speed and assists in cooling the engine. Like the flywheels found on most working engines, from automobiles to tractors, a flywheel uses inertia to smooth crankshaft rotation, lessening vibrations from the engine and lessening main bearing wear.

How Big a Chainsaw?

For felling tall trees, cutting firewood, or just clearing scrub brush, a chainsaw whittles even daunting tasks down to size, but it's vital to match the saw to the job you intend for it. "Power matters when choosing the right saw," says Tim Ard of Forest Applications Training, Inc. An estimate of what a given machine is *capable* of can be made from its engine's cubic-centimeter displacement, but size isn't the only factor to consider.

It has been said about Americans that we associate bigger with better, and in many cases, like motorcycles and off-road vehicles, that indeed seems to be the case. But bigger isn't always better, or even good. Nor is it entirely dependent on a user's physical size, muscle strength, or gender. It *does* depend entirely on the user.

That is demonstrated by my wife, whose favorite chainsaw is a 50cc model with a 20-inch bar. How big a chainsaw is too big has nothing to do with gender, or—presuming you possess normal strength—with how much muscle you have.

Meanwhile, a relative, a rather burly man, is missing his four front teeth and a piece of bone from his chin because he tried to operate a much less powerful saw with one hand when he was sixteen, showing off to some friends.

NEVER TRY TO OPERATE A CHAINSAW WITH ONE HAND.

There are two handles on every chainsaw, and both an operator's hands should be firmly wrapped around both of them all the time.

How Long a Bar?

A chain bar is a guide, a piece of laminated steel with a groove running longitudinally around its circumference; the groove contains and guides the chain's underside drive teeth. A small drive sprocket located at the side of the engine is essentially an extension to the engine's crankshaft, and provides unadulterated power to the teeth. A second, larger star gear (sprocket) at the end radius of a saw is a bogey wheel.

It's a common practice to try and get more from a chainsaw by fitting it with a longer bar than it was intended to use. This doesn't harm the saw directly, but does cause an engine to work harder, and by making it turn more chain links, you're actually subtracting from its horsepower. If your saw was designed to operate a 14-inch bar, don't fit it with a 16-inch bar. Likewise, it detracts from the ability of your saw to fit it with a smaller bar and chain than it was designed to use.

That said, a shorter bar is easier to handle. It's less likely to kick back uncontrollably—partly because shorter bars are found on less powerful saws. That works out, so long as a sawyer's ego doesn't override sensibility and capability. You'll know when it's time to step up, and the sometimes hard-won information on these pages can help to quicken that process.

Laminated Bar or Solid?

This is a common argument. Some chainsaw guide bars are milled from a solid piece of steel, while some are made from three layers, spot-welded together in a steel sandwich.

Solid-steel bars are generally more expensive to manufacture, as they must be precision milled, then hardened or "tempered." Ideally, a bar can flex without breaking or chipping—a characteristic of softer steels—yet be hard enough where it contacts its chain to offer minimal friction and resist wear without becoming brittle enough to chip.

Most are zone-hardened, meaning that the track into which a chain's drive teeth sit is made harder than the interior of the bar to lessen friction against its chain while maximizing its overall toughness (a bar's ability to flex without breaking). Sprockets are built to be rebuilt, with precision needle-bearings. Solid bars are the choice of chainsaw sculptors who subject their bars and chains to twists, turns, and stresses that might cause laminated bars to separate.

It began as a log; now it's a bear.

A chainsaw-sculpted eagle.

A chainsaw-carved seat, worthy of anyone's seat.

An ornately carved seat of pine.

Chainsaw sculptors turn logs into works of art, using modified bars and chains, and techniques that show the capabilities of the tool (but are not recommended for average wood cutters).

A chainsaw-carved snowy owl.

Laminated bars are spot-welded together, with a hardened inner plate of steel laminated to softer plates on either side. This arrangement allows for minimal friction and wear against a rotating chain while providing the toughness needed to survive stress against the bar. Laminated bars are less expensive, but that shouldn't be associated with low quality, as they seem to be the choice of most sawyers—from homeowners to professional foresters—people who do not need to twist and turn their chainsaws, only to cut with them.

While a debate goes on as to which type of bar is better, all of them must meet minimal safety and quality standards set by the American National Standards Institute (ANSI). Any real difference in quality or weight is imaginary for typical sawyers. Either will serve equally well.

The Right Fuel

Gasoline-powered chainsaws have 2-stroke engines that cannot operate on straight gasoline. They run on what used to be known as "Power Mix," a gasoline blended with oil at a ratio of 50:1; some gasoline stations had a dedicated pump that dispensed nothing else. Outboard boat motors, dirt bikes, and many other 2-cycle applications couldn't operate on straight gasoline without damage to valves, pistons, and maybe cylinders.

Chainsaws are one appliance that still requires blended gasoline. Sometimes you can purchase the fuel pre-blended in 1-quart bottles (currently, prices may well change) for about $7 per container.

But very often, pre-mixed fuel is unavailable, especially (almost ironically) in places where chainsaws are used the most. And, in any case, it's less expensive to mix your own fuel. Chainsaw mix oil is typically sold in 2.6-ounce containers for about $3 each or in six-packs of these containers for around $16. The proper oil reduces internal friction, lowers engine temperatures, and helps to reduce smoke from exhaust emissions. Most contain synthetic friction-fighters. In times past, it was common to just use motor oil, but this should not be done unless absolutely necessary.

While blending gasoline and oil to make power mix isn't a precision operation, care should be taken to ensure a ratio that's as close to 1 part oil, 50 parts gasoline as possible. Too much oil—or the wrong oil—and you'll foul your spark plug with unburned carbons prematurely. Too much gasoline and your engine will run hot, usually noticeable as a burning-off of a spark plug's electrodes.

Some manufacturers recommend a 40:1 ratio of gasoline to oil instead of 50:1. In the real world, professional loggers save themselves a headache and just mix one 2.6-ounce bottle of oil into 1 gallon of the correct grade of gasoline. I, of course, do not suggest doing anything except what is recommended by a saw's manufacturer, but one bottle of oil to one gallon of gas is a rule of thumb that everyone I know goes by.

When mixing your own fuel, it isn't necessary to use a premium- or Super-grade gasoline (91–93 octane). However, chainsaw manufacturers recommend against using the lowest grade of 87-octane. The most preferred grade is mid-range 89 octane.

1 gallon gasoline: 2.6 ounces of oil.
2.5 gallons gasoline: 6.4 ounces of oil
5 gallons gasoline: 12.8 ounces oil

Ethanol

The Environmental Protection Agency (EPA) has approved gasoline with 15 percent ethanol in automobiles made since 2001, but mandates against the use of either E10 (10%) or E15 (15%) mixtures of alcohol/gasoline in any 2-stroke engines. (Non-ethanol fuel is E-Zero, or E0.) According to a Department of Energy study, ethanol-mix fuel causes hotter operating temperatures, misfires, and shorter life of internal parts due to decreased lubrication.

Marv Klowak, Global VP of Research & Development for Briggs & Stratton, America's largest manufacturer of small engines, states, "Ethanol has inherent properties that can cause corrosion of metal parts, including carburetors, degradation of plastic and rubber components, harder starting, and reduced engine life. The higher the ethanol content, the more acute the effects." According to the Outdoor Power Equipment Institute, the industry's trade group, service centers concur with that statement.

Bottom line: Do not use ethanol-mix fuel in your chainsaw. If you must use ethanol gasoline, treat the fuel mixture with a protectant additive, like *Sta-Bil*®.

Gasoline Stabilizers

Gasoline evaporates quickly, leaving behind a "varnish" that can foul spark plugs and coat carburetor needle valves, effectively changing their adjustment. The problem is a common one that occurs naturally with the effects of time and use, but it's especially critical when saws are used infrequently or stored for long periods of time.

Some authorities recommend emptying all of the gasoline from a saw's tank and fuel lines before putting it into storage. Most people do not do this. However, it is smart to add a "stabilizer" to the fuel tank, as they help to keep gasoline in solution inside a sealed fuel tank for up to twelve months.

There are several fuel stabilizers on the market, but the brand that is recommended most often by service technicians is *Sea Foam*®, available at most auto parts supply stores.

Use the Right Container

While any approved (federally mandated to be red-colored) container can hold fuel, a gas can that's made for chainsaws generally holds 2.0 to 2.5 gallons, has a spill-proof pouring spout that can be hermetically sealed, and a screw-cap vent on the end opposite the pour spot. The vent is especially essential on warm days, when you should use it to relieve internal pressure that could build high enough to rupture the container.

Avoid the common mistake of mixing too much fuel. A gallon at a time is more than sufficient for most folks. Untreated fuel can be stored for three months without treatment, but is sometimes stored for a year, with appropriate fuel additives. You should always swirl gas around inside the sealed container to mix it, before fueling your saw.

When Fueling

There is no fuel gauge on a chainsaw, so check its fuel and bar oil reservoirs before every use—it's surprisingly common (and slightly embarrassing) for an operator to be unable to start his chainsaw because the machine is out of gas.

Place your chainsaw on the ground, with fuel cap facing upward. Open the cap on the fuel tank slowly, especially on hot days, to bleed off internal pressure. Clean the fuel cap and the area around it to ensure that no sawdust or other debris falls into the tank.

Cutting wood is hard work, and a chainsaw's bar needs to be lubricated to keep it running smoothly.

Chainsaw fueling cans are traditionally smaller than others, with a volume that equals multiples of oil bottles, so that you can create Power-Mix for 2-stroke engines that require it simply by dumping in 1 or 2 bottles of oil, then topping off the container with gasoline.

Oil Reservoir

Even more important than fuel is keeping the oil reservoir full—running out of gas won't do any harm, but running out of lubricating bar oil can damage your saw's bar and chain, and overwork its engine.

Good ol' boys have long tended to just reuse 30-weight motor oil drained from their car's engine when they changed the oil, as bar and chain lubricant. Mention of that practice made my chainsaw expert, Pete, grimace.

Lubrication is essential for the bar and chain of a chainsaw, so always keep the oil reservoir full when using your saw, preferably with biodegradable vegetable oil. (Most saws leak oil during long periods of disuse, and should be stored in a case, with oil reservoir empty.)

While it is true that used motor oil is thinner and slings from a rotating chain more easily, and that used motor oil contains microscopic metal impurities that aren't conducive to chain life, the argument over whether clean motor oil isn't as good purpose-made bar oil is far from conclusive. Chainsaw dealers insist that purpose-made bar lubricant is superior to motor oil, but that might be factory training seminars talking. In fact, even the USDA Forest Service isn't anxious to offer an opinion on this point.

What the Forest Service does have an opinion about, however, is that petroleum-based bar-and-chain oils aren't good for the environment, regardless. According to a Forest Service study, "Every year thousands of gallons of chain-and-bar oil are carried into the forests and none returns."

In lieu of spraying harmful oils around the woods, authorities suggest using "green" vegetable-based bar-and-chain oils, like Ultra Lube or Green Earth Technologies oils. The price of either vegetable- or petroleum-based oils is roughly equivalent and, with internet shipping, both are readily available, so there's no reason not to be environmentally conscious.

Be aware that it's normal for small amounts of oil to leak from the oiler when the saw isn't running. I personally don't mind; my saws hang from hooks above a wooden workbench and the oil is good for preserving the surface. But leaking bar oil onto the wrong surfaces can be harmful to a relationship.

Chains

The Importance of a Tight Chain

Our chainsaw expert says, "It's important to keep that chain tight." Chainsaw sculptors sometimes purposely run a loose chain on some of their specialty saws, but for wood-cutters, a snug chain is the Order of the Day.

A tight chain on a chainsaw is very important, this chain is dangerously loose.

Chains can loosen on their own during the course of ordinary work, but the big reason for a loose chain is that it stretches (actually becoming longer under duress), and the problem is especially chronic with new chains that haven't been broken in yet. A loose chain can jump its track in the bar, and while every saw is equipped with a Chain-Catcher, whose job it is to bunch up a chain that does that, serious lacerations can occur. For that reason, work-grade gloves should always be worn not only when working with a saw, but when working with a chain.

Also worth mentioning is an instance in which I purchased a brand-new orange-colored saw—arguably one of the best brands made—from a farm-supply chain store. I could not keep a chain tight on that saw, especially when cutting the jack pines that are notorious not only for gnarly grain and iron-hard knots, but for being killed by the Jack Pine Budworm (*Choristoneura pinus*) when they're 50 feet tall. After ruining two bars and chains when the chain suddenly whipped free at speed, I took the saw to my chainsaw expert, Pete. He replaced its star-type drive sprocket with the clutch type drive that it should've had in the first place, and I haven't had a problem in three years since.

Keeping a snug chain is necessary, because a chain that jumps its track can damage itself and the bar, and injure a sawyer. Modern saws have "toolless" tightening mechanisms, but for all other saws, loosen the two bar-retaining nuts, turn the adjustment screw clockwise until the chain is snug but rotates freely by hand, then re-tighten the nuts.

The Jack Pine Budworm is one of the numerous native and invasive insects that make it necessary to cut trees before they fall onto something of value—this budworm only kills large, adult trees. (Photo courtesy USDA.)

Adjusting the Chain

Many older saws are equipped with adjuster screws. First, set the chainsaw on a flat, stable work surface, and be sure that the ignition is in the OFF position; if feasible, pull the spark plug wire from the plug.

Loosen the clutch cover/bar nuts—slightly, a quarter-turn is usually enough to allow the bar to slide forward or back.

Using the proper-size (usually flat-head) screwdriver, turn the chain tension screw clockwise until the slack is gone from the chain. Tension screws are usually located on either the front or side of the chainsaw body.

When the chain is snug, re-tighten the clutch cover nuts.

Some newer saws have a very convenient, very fast chain tensioner that is a crank-type. Just fold out the crank handle, and rotate it counterclockwise to fully loosen the chain, and then use the thumb-wheel to tighten the chain, and lock the handle back into its folded position.

As important as having a tight chain is to not have it too tight. A too-tight chain causes premature wear of bar, chain, and nose sprocket. Tractor Supply Company® suggests a "Snap Test," in which you grasp the top of a chain between thumb and forefinger, then pull it away from its bar; a properly tight chain snaps back into position. In any case, none of the drive teeth should completely clear the bar's groove when you pull the chain outward, but the chain should rotate easily when turned by hand.

Type of Chain

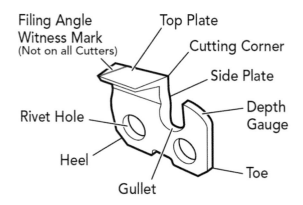

The terminology for the parts of a chainsaw cutting tooth; the cutting corner, adjacent to the top plate and the side plate is the part that gets sharpened; the depth gauge doesn't cut, but "rakes" severed wood out of the way, while serving as an indicator of chain wear. The cutting surface must rise above the raker/depth gauge, and if a chain has been sharpened too many times, it will not.

A chainsaw's chain is comprised of five parts. First, there are left and right cutting teeth, positioned facing the center on either side. Sited between the cutters and facing downward, a guide bar's slot, are drive teeth. Cutting and drive teeth are linked together by tie straps, and all of these are fastened together with flanged rivets that act as hinges. Very much like a bicycle chain that bites.

Several types of chain exist, from abrasive diamond concrete-cutting types to very hard carbide-toothed cutting chains used by rescuers to cut through storm-smashed buildings, where timbers might contain nails and other foreign objects.

Chainsaw technician Pete is adamant that both homeowners and professional loggers are best advised to stick with an ordinary high-speed steel, chisel-toothed chain. This design is the most field-proven cutting configuration. Chains need to be sharpened and even super-hard carbide-toothed chains dull eventually, but they're almost impossible to re-sharpen.

Keep Your Chain Sharp

Sharpening a chain not only helps a saw to cut better, it extends service life of a chain by decreasing heat and friction, as well as decreasing demand on the saw's engine. And, like a sharp knife (or any cutting tool, for that matter), a sharp saw is safer to operate, as it requires less pressure against the wood it's cutting.

Cutting teeth have two important features:

Top Plate Angle

Looking down on a cutting tooth, the top plate angle is the horizontal angle of the cutting tooth. Put simply, this is the cutting edge of a tooth, the edge you sharpen. It may be factory set up to 35 degrees, depending on the chain.

Tilt Angle

Looking on a cutting tooth from above, the tilt angle is the vertical angle at which the uppermost flat side—the blade, onto which the cutting edge is filed—of a tooth is slanted. On the majority of chains, the tilt angle is 90 degrees (perpendicular, or flat) to the chainsaw bar.

When facing the flat side of a chainsaw bar, half of the teeth on the chain face inward and forward from the left side, half of them face the opposite direction.

Because teeth on a chain alternate their facing sides every other tooth, teeth must always be sharpened from the *inside* face of the tooth *outward*. Sharpening should be performed on every other tooth for each side of a chain; first from one side, then from the other—first you sharpen the left side, then you sharpen the right (or vice versa).

File Size

Chainsaw bars come in an array of lengths, widths, and types that accommodate a range of needs, but be sure that, if yo ever need a new one, that ou get the proper bar, and chain, as recommended by the manufacturer of your saw.

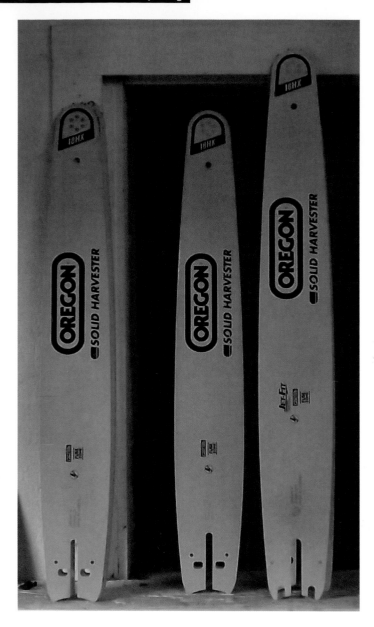

Different bars for different jobs.

As the size of a chain's cutting tooth increases, so does the radius of the correct size file to match it. Information on the correct size sharpening file for your chain is available from the chain manufacturer.

That being said, it's pretty easy to visually determine if your file size is correct. When you place the file into a tooth's radius, it should fill that radius almost entirely. Some skilled sharpeners can do a pretty fair job with a file that is actually smaller than recommended, but for best (and safest—an improperly sharpened chain tends to kick back more) results, always use the correct size file:

3/8" PICCO chain is sharpened with a 5/32" round file.
.325" chain is sharpened with a 3/16" round file.
3/8" chain is sharpened with a 13/64" round file.

Freehand Sharpening

Sharpening a Chain with a Round File

This is the most common method of sharpening a dull cutting chain. At the same time, it's the least recommended among professionals.

But a chain, like any cutting tool, gets dull with use, and the job of a working chainsaw is demanding. A round file is (arguably—much depends on the skill of the user) the least accurate sharpening method, but it continues to be popular because the file is the only piece of equipment needed. That said, always use a sharpening guide.

To sharpen a chain with a round file, the file must be held parallel to each tooth's tilt and top plate angles; this requires that the tilt and top plate angles be maintained *visually*. This is nearly impossible, even for experts, and is possible to destroy the chain.

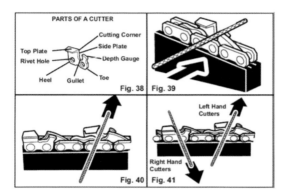

To Sharpen a Chain *without* an Angle Guide

1. Place the saw in a vise, if possible
Clamp the jaws of a table vise around the chain bar to stabilize the saw during sharpening. It is possible for experienced sawyers to sharpen chains in the woods, without a fixture or vise, but it's also possible to destroy your chain.

2. Position the file on a tooth
Place the file on an inside-facing tooth and set it so that it is parallel to the tooth's tilt and top plate angles.

3. File the tooth
Stroke the file away from your body—not back-and-forth—while maintaining the correct angle. Direction of strokes should be from the inside face of the tooth towards its outside face.

4. Proceed to other teeth on that side
Follow the same procedure for each tooth facing that direction, on that side of the chain. Advance the chain by hand to bring each tooth into filing position.

5. Sharpen the other side
The saw can either be flipped around in the vise or the sharpener can walk around to the other side of the saw. Always sharpen cutting teeth from bar to outside.

6. File each tooth's raker height with a raker gauge and flat file
This last step is the same for all methods of sharpening. The raker height of each tooth must be filed with a flat file. A raker file gauge makes doing this fast and easy, since the gauge protects the sharp edge of the tooth from the file.
   Sharpening raker teeth with an unguided round file is a *last resort*.

Sharpening a Chain *with* an Angle Guide

Using a *file guide* makes sharpening with a round file more accurate. File guides have markings on them that allow you to control the file's tilt and top plate angles, and they allow quicker sharpening.

1. Place the saw in a vise
Clamp the bar of the chainsaw in a vise to stabilize it while sharpening.

2. Position the file and holder on a tooth
Set the file and its guide over an inside-facing tooth on the chain. *Use the markings on the file guide to visually line up the file at the correct angle for the tooth.*

3. File the tooth
Stroke the file against the tooth *away from yourself,* from the inside face of the tooth towards its outside face.

4. File the other teeth on that side
Follow the same procedure for each cutting tooth on that side of the chain. Advance the chain by hand to position the next tooth.

5. Sharpen the other side

6. File each tooth's raker height with a raker gauge and flat file

Sharpening a Chain with a Sharpening Jig

A chainsaw Sharpening Jig is a fixturing tool that doesn't allow you to do it wrong. Jigs are more expensive than other sharpening methods, but having one may be worth the investment, as they ensure a precisely sharpened chain in the shortest time, with the least amount of effort.

1. Place the saw in a vise
Place the bar of the chainsaw in a vise to stabilize it while sharpening.

2. Prepare the jig
A typical jig fits onto the saw's bar impeding rotation of the chain, and has adjustments for aligning the correct depth, top plate angle, and tilt angle.

3. Sharpen the first tooth
File inside to out, in one direction; this helps to extend the life of round files. Sharpening with a jig is fast, so take care not to over-sharpen; about ten strokes will sharpen a very dull chain.

4. Proceed to the next tooth
Advance the chain by hand; the jig will allow the chain to rotate freely.

5. Sharpen teeth on the other side
Remove the jig and reposition it to other side (if necessary) and repeat the process.

6. File each tooth's raker height with a raker gauge and flat file

Replacing the Chain

Sooner or later, you're going to have to put a new chain onto your chainsaw's bar. This is a relatively simple process, and experienced sawyers usually carry a spare, sharpened cutting chain in their "possible" tool box so that it can be quickly changed in the woods.

To begin, loosen the two locking nuts that hold the bar in place—this can be accomplished with the appropriate size wrench or socket, but most chainsaws come with a combination tool—called a *scrench*—that incorporates a bar socket, a spark plug socket, and a flat screwdriver for adjusting chain tension—all the tools that you'll most likely need to service your chainsaw in the field. Remove the nuts and the chain cover, exposing the entire length of the bar and chain. Be aware that the bar will come loose and tip downward when restraining nuts are removed, and the chain will probably wrench free of the bar's track of its own accord.

Slide the guide bar rearward until the chain is loose. The bar will come right off with the retaining nuts removed. If the bar does come off, there's no harm done, and it cannot be put back on the wrong way, but it needs to be held in place over the retainer nut studs with hand pressure when replacing a cutting chain.

A danger that you do need to be aware of is that the new cutting chain can be installed backwards—this is a little embarrassing, but it happens all the time. If a chain is installed backwards, there could be damage to the saw's drive components, but probably not before you discover your mistake. Most notably, the saw won't cut—and you'll notice that right away.

Avoiding putting a new chain on backwards is simple. Forget the proper nomenclature; just remember that, looking down onto the top of the guide bar, the angle of all of a chain's cutting teeth should point forward, toward the end of the bar.

When the new chain is in place, the cutting chain is on correctly, and its drive teeth are seated fully into the guide slot of the bar, replace the cover and snug-down both retaining nuts hand-tight.

With the chain-tightening screw turned fully in, that is, so it holds the chain at it loosest, pull the bar out—away from the saw, until its chain is snug. It's important to have the tightness-adjuster screw at its maximum looseness setting (if this is applicable) so that you'll get as much travel from it as possible, as the chain naturally stretches and loosens from wear.

Replacing the Bar

If you can replace a chain, you can replace a bar. The big difference is that, in this case, you completely remove a bar's retaining nuts and take bar and chain off entirely. Replace old bar with new bar, and then follow the same procedure that you do when installing a new cutting chain. Make sure that the new chain is properly seated in the bar's guide groove and that it rotates smoothly (if your saw has a chain brake, it must be disengaged to move the chain—and, of course, the saw *must not* be running).

Repairing a Bar

Repairing a bar is sometimes necessary whenever a chain jumps its track, especially with more powerful saws, having motors of 50cc or greater. If a chain jumps its track, the drive teeth can deform and burr-up the groove that guides them, sometimes to the point of preventing drive teeth from being re-inserted or from sliding smoothly.

My chainsaw expert, Pete, repairs bars that have been damaged thusly with a small mill file that will fit into a bar's guide groove. Lightly filing the groove from inside removes burrs and permits chains to glide as smoothly as if the bar were new.

If a bar is bent or twisted, it's best to replace it. The force that must be exerted to straighten such damage will probably only do more damage.

Replacing the Spark Plug

Without a spark plug, your chainsaw's engine cannot run. The same is true if our saw's spark plug isn't "firing" (sparking) as it's supposed to, when it's supposed to.

None of this is probably a revelation to most homeowners, who tinker with their own lawnmowers and other small-engine appliances, even though it's entirely too common for operators to neglect the condition of these critical components. Remove and check the plug routinely to see if its gap (the distance between electrodes) is correct, to inspect electrodes for uneven wear, and to clean them of fouling.

If a spark plug seems to be fouling a lot, that's an indication that your saw's engine is not burning fuel cleanly. Causes might include too much oil in the fuel mix, causing it to burn incompletely, and leaving oily residue on a plug's electrodes. Or your saw might be fitted with the wrong spark plug. Or the engine's compression rings might be worn. In all three cases, a saw will likely smoke excessively and misfire.

A fouled plug can be cleaned and re-gapped, but using too "hot" a plug will cause its electrodes to burn unevenly and prematurely. This problem can be remedied by filing and re-gapping only until electrodes are burned away. A saw with too hot a spark plug may otherwise appear to be running okay, at least until the gap between electrodes becomes too wide.

To help prevent the above problems, never replace your saw's spark plug with any other than the model recommended in its owner's manual. Some shadetree mechanics might suggest going to a hotter or cooler spark plug, but you can generally rely on the engineers who designed your chainsaw to know which spark plug works best in it.

Operating a Chainsaw

"Possibly the most useful safety resource arrives boxed with a saw," according to Tim Ard, owner of Forest Applications Training, Inc. "First thing is to just read the manual."

Having grown up in timber country, where forests are considered to be crops, I've seen more chainsaw injuries than I'd like to remember. Chainsaws are serious tools, and, although I promised to keep the tone of this book light, operating a chainsaw is a deadly serious proposition.

Do not rely on any safety device. Pretend your saw doesn't have a chain brake, and act like the bar is a laser, capable of cutting off any portion of your anatomy at a touch. Do not let chainsaw-resistant garments permit your mind any lapse in security-consciousness.

Do not operate a chainsaw without a firm, two-handed grip. This cannot be repeated annoyingly enough—it should be tattooed on every operator's brain. I had a brother-in-law who should have known better by sixteen years of age, but became overwhelmed by the teenaged boy's penchant for showing off in front of friends. He was operating a small Remington "pocket saw," and, is so often the case, he mistook smaller for safer.

The little Remington kicked back smartly, and, partly *because* of its light weight, its speed took him entirely by surprise. The radius of the bar caught the left corner of his mouth, taking out two teeth—two on top, two on the bottom—removing a goodly chunk of both lips and leaving a significant cleft in his chin bone. Never attempt to operate any chain saw with one hand, no matter how large or small.

Bucking a downed jack pine.

According to OSHA and Forbes magazine, lumberjacking is the most dangerous occupation in the world; do everything you can to prove them wrong.

Bucking a mostly-felled tree.

Situational awareness - nowhere does that catchphrase have more genuine meaning than when you're lumberjacking.

Gripping a Chainsaw

I'll readily admit to operating a chainsaw backward—I'm left-handed, and I grew up using right-handed saws. That means that I hold the bar handle at the front of the saw in my right hand while I yank the pull starter on the left side of the saw with my left, strongest arm. Then, when the saw is running, I grasp the forward bar handle with my left hand, remove my right hand from the front handle, and place it on the trigger (throttle) handle.

When I'm operating the saw, my left (strongest) arm is on the forward handle, where I have maximum strength against kickbacks. According to OSHA, this is the proper grip for right-handers, but that's merely a coincidence in my case. If I were operating a left-handed chainsaw, like a southpaw is supposed to, I'd be doing it "wrong." Fortunately for me, left-handed saws are as rare as turtle feathers, so I've learned to use right-handed chainsaws.

My wife, who is mostly right-handed (she's actually ambidextrous), starts her saw, then grips the forward handle with her right, strongest arm, operating the trigger with her left hand. According to OSHA, this is backward. But, like me, she feels safest with her strongest arm guarding against kickback.

Our litigation-happy society prohibits me from contradicting the experts, but I want to pull the starter cord straight back, not across the saw, as recommended. And, whether I was right or left-handed, I'd want my strongest arm gripping the saw's forward bar handle to prevent the bar from kicking back against any part of my body.

SAFETY GEAR

Chainsaw Chaps

Since 1965, the US Department of Agriculture Forest Service has provided cut-resistant, protective chaps for chain saw operators, and they've prevented innumerable serious injuries (most chainsaw injuries are very serious). Data collected by the Missoula Technology and Development Center (MTDC) tracked chain contact injuries for more than thirty-five years, and resulted in many improvements.

The original Forest Service chain saw chaps incorporated four layers of ballistic nylon and withstood a chain speed of 1,800 feet per minute without penetration.

In 1981, Forest Service chaps were redesigned to be stronger and more comfortable. A Kevlar pad was added below the ballistic nylon shell, increasing the protective rating to a chain speed of 2,500 feet per minute, while cutting weight by 40 percent.

In 2000, chain saw chaps were again improved to protect a larger area. Five layers of Kevlar enabled them to withstand chain speeds of 3,200 feet per minute without cutting through. Because most chain saws are operated by right-handed people, the majority of contact injuries occur to the left leg, so coverage on the left side of the left leg was increased by 2.5 inches and on the left side of the right leg by 1.5 inches Weight of the chaps increased 6 to 8 ounces, depending on the size.

Today, chainsaw chaps are more effective, offering more protection than ever. At around $70 a pair, there is no excuse for not having them.

How Forest Service Chain Saw Chaps Protect

Chain saw chaps don't actually resist the cutting force of a chain—not many things can.

In chaps used by the Forestry Service, a heavy-Denier (Denier is the number of threads per inch) shell of Polyurethane-coated ballistic nylon covers a 5-layer Kevlar pad. The nylon shell resists oil, routine abrasion, and other phenomena that might damage or reduce the effectiveness of Kevlar.

The Kevlar pad is a pillow of waterproof polyurethane-coated nylon duck surrounding five layers of Kevlar: namely, woven Kevlar, felted Kevlar, woven Kevlar, woven Kevlar, and felted Kevlar. Kevlar is an *aramid* fiber, similar to the more supple Nomex gloves worn by fighter pilots, but with even higher flame-resistance (FR) properties. When exposed to temperatures higher than 500 degrees Fahrenheit, the nylon shell melts, but the Kevlar doesn't.

But chainsaw operators have little concern over a chaps' ability to withstand heat. What they care about is that Kevlar fibers caught by a running chain will bunch up and jam a saw's drive sprocket, usually stalling its engine but stopping a chain from moving in any case.

Properly fitted chaps, in the correct size, maximize the protection they provide. Chaps should be adjusted for a snug, comfortable fit that keeps them positioned correctly on the legs. Properly sized chaps reach 2 inches below the boot tops.

Like a bullet-resistant vest made from Kevlar, chain saw chaps aren't made to do anything beyond preventing a catastrophic injury. According to published user reviews, some folks believe that their chaps make them immune to injury from their chainsaws. That is simply not true. Never, ever rely on your chaps to save you from poor operator techniques. Always treat the bar of a chainsaw as if it were a *Star Wars* light-saber, capable of removing a limb at the merest touch—that isn't so very far from the truth.

Care and Cleaning of Chainsaw Chaps

The Forestry Service regards chainsaw chaps as critical safety equipment and recommends treating them accordingly. According to that Service's literature, "correct and timely cleaning reduces general wear and tear and the chaps' flammability." (It is possible that gasoline-soaked chaps might catch fire, in which case, the flame-resistant properties of Kevlar might, indeed, be an important trait, but generally, a chainsaw user is concerned only with a chap's ability to stop a running chain.)

The Service endorses regularly cleaning chain saw chaps with a citrus-based cleaner called CitroSqueeze, which has been approved by DuPont for cleaning both Nomex and Kevlar aramid fibers.

Never machine wash or machine dry chain saw chaps. Use a garden hose and nylon-bristle brush to remove dirt. For heavy soiling, soak chaps overnight (for at least four hours) in a properly diluted solution of warm water and CitroSqueeze. After soaking, scrub with a nylon-bristle brush, rinse thoroughly in cold water, and air dry.

For light cleaning, use CitroSqueeze solution in a spray bottle containing 1 part CitroSqueeze concentrate to 10 parts water. Spray soiled areas, and scrub with a bristle brush. After thirty minutes, lightly scrub, then hose chaps with cold water. Line dry.

Repairing Chainsaw Chaps

Repair cuts and holes in chap's outer shell immediately to keep debris from the inner Kevlar layer. Clean chaps with a damp cloth and let dry.

The US Forestry Service recommends a product called Seam Grip® Repair Adhesive and Sealant, from McNett Corporation. This rubber-based paste is flexible, stretchable to 100 percent, waterproof, and abrasion-resistant.

To repair rips up to one-half inch, simply apply a dot of Seam Grip over the hole and allow it to dry. It's important that you *do not* get adhesive onto the Kevlar pad, which might restrict the free movement that makes fibers bunch against a chain.

To repair tears longer than one-half inch, slip a piece of writing paper, about 2 inches larger than the hole on all sides, inside the chaps, between Kevlar layer and outer layer.

Lay the area being repaired on a flat, hard surface and apply a smear Seam Grip directly onto the repair below the damaged area. Press the nylon shell down onto the paper to work the adhesive into the weave of the nylon shell.

Squeeze Seam Grip onto the outside of the tear, covering all sides. I like to work adhesive into the outer weave thoroughly with a popsicle twig, a popsicle stick, or even a forefinger tip.

Allow the chaps to dry for twelve hours before use.

Inspection and Retirement

Chain saw chaps need to be inspected frequently and replaced when necessary. Signs that your chaps need to be retired include:

When the outer shell has accumulated numerous holes and cuts; holes permit bar oil and other contaminants to weaken Kevlar pads. Oil has adhesive properties that attract and hold sand, sawdust, and dirt, reducing the level of protection.

If improper repairs have been made, this also reduces the level of protection. Machine or hand stitching through a Kevlar pad effectively binds individual fibers together, preventing them from moving, and bunching up like they should.

Remember, chaps do wear out, like any piece of working equipment, and when they're worn, they should be replaced.

For more, and the latest, information about chaps, contact:
USDA FS, Missoula Technology and Development Center
5785 Hwy. 10 West
Missoula, MT 59808-9361
Phone: 406-329-3978
E-mail: wo_mtdc_pubs@fs.fed.us
Website: www.fs.fed.us/t-d

CitroSqueeze is available from:
CDR Chemical, Inc.
16182 Gothard St., Suite J
Huntington Beach, CA 92647

Phone: 888-270-4237

Website: www.cdrchemical.com

Seam Grip is available from numerous outdoor retailers.

To learn of the retailers close to you, contact:

McNett Corporation

1411 Meador Ave.

Bellingham, WA 98229

Phone: 360-671-2227

Fax: 360-671-4521

Website: www.mcnett.com

Headgear

All-in-One Headgear

Unheard of when I was a boy, cutting up to forty cords of firewood each fall, next-generation AIO headgear has become as must-have as a chainsaw when cutting wood, especially when felling trees.

I hope that my readers will forgive me here, but I'm going to come right out and recommend the Peltor System from 3M. It's what my wife (that's right, guys, she has her own lumberjacking gear, including a 40cc chainsaw) and I both use.

Although they don't actually manufacture them themselves, All-in-One kits, consisting of chainsaw chaps, hardhat, ear protection, eye protection, and more, are offered by most chainsaw makers, to help ensure that their customers are safely attired.

Hardhat

The new PELTOR all-in-one head protection system is an inexpensive accessory that every lumberjack, regardless of skill level, should wear whenever cutting trees.

3M's Peltor system, which is sold under a lot of chainsaw manufacturer's brands, with their logo on it, begin with an ANSI-approved, high-density polyethylene (HDPE), Blaze Orange hardhat with a "strongback" crest molded into its crown. Four-point ratchet or pin-and-hole suspension systems with built-in sweat bands make the hardhat compatible to every head and comfortable enough to wear all day. A molded-in brow visor keeps rain, sun, and falling debris out of your eyes.

Several tons of wood deserves your respect, and if you don't give it willingly, it will eventually take it from you.

This Peltor (R) all-in-one headgear from 3M is a state-of-the-art protection that no lumberjack should overlook.

Visor

A pull-down visor of steel-mesh supported in a stout perimeter frame provides full-face protection from flying wood chips, debris, and even insects—many a lumberjack has been forced to flinch mid-swing by a blackfly that suddenly dove into his eyeball. Vision is unrestricted; you won't even notice the mesh in front of your eyes, but you will notice that it diffuses bright sunlight as if you were wearing a pair of sunglasses (which can, in fact, be worn comfortably behind the mesh visor). Unlike conventional solid see-through shields of Lexan® or Plexiglass, the mesh offers lots of ventilation for those hot, sweaty days in the woods, and it cannot fog up.

Although the mesh visor is probably adequate by itself, 3M recommends that additional eye protection be worn under it, just to be sure. Not ironically, 3M makes a line of athlete-influenced safety glasses that have proved ultra-light and comfortable. Wrap-around lenses are made from tough polycarbonate, treated with an anti-fog coating, and rated to keep out 99.9 percent of potentially harmful Ultraviolet rays. Some have a 2x diopter to assist folks who need reading glasses. A foam gasket around the lenses keeps out nuisance dust and insects. Priced at around $8, I've actually forgotten that I was wearing them.

Ear Protection

Completing the Peltor's all-around protectiveness are a pair of sound-absorbing ear muffs. Hearing protection has become accepted by typically hard-nosed timbermen over the past couple of generations. A chainsaw with a good muffler emits 110 decibels. By way of comparison, a rock concert puts out 105 decibels, and is right on the edge of what the National Institute of Occupational Safety and Health (NIOSH) considers permanently harmful to a listener's hearing—when I was a kid, you could identify folks who went to rock concerts regularly, because they said, "What?" a lot.

The Peltor system incorporates individual ear muffs that snap onto either side of its hardhat, then swivel up, held in place against the hardhat by spring tension when not in use. The heavily cushioned muff is able to acoustically seal over most beards or sideburns. The muffs seal out hard, jarring noises, yet allow you to hear conversation.

NIOSH says ". . . the best hearing protector is the one that is comfortable and convenient and that you will wear every time you are in an environment with hazardous noise." If heavy beard or sideburns prevent earmuffs from getting a good seal, there are several makes of OSHA-approved ear plugs that can be used instead.

Simplest, most inexpensive, and, in my opinion, best of these for all-around use are soft foam models that are rolled and compressed between thumb and forefinger, then swell up again when placed into an ear canal. These are typically sold several pairs in a package, for less than 25 cents a pair. Some folks even wear foam ear plugs under earmuff protectors, but I think these impair your hearing too much—you need to be aware of what's going on around you.

AXES

Anatomy of an Axe

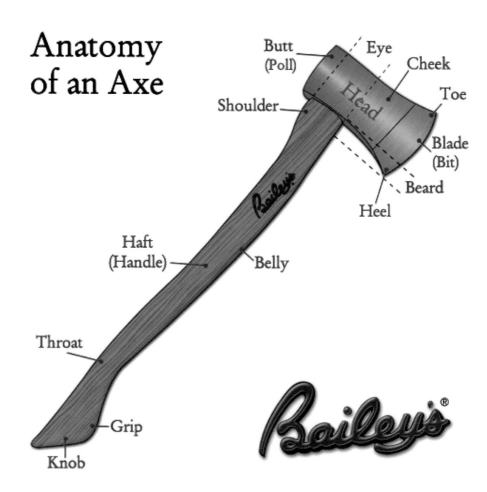

Anatomy of a typical single-bit (single-edge) axe. (Courtesy BAILEY'S Axe Company).

It's no coincidence that the battle axe of lore is associated with so-called barbarians, like the Vikings and Celts, and on over to Russia. Northern woodsmen who also used their axes to cut trees from the forests around their chosen homes to build, and, when necessary defend (or expand, by way of territorial conquests) those homes used them the way an accountant uses a calculator—namely, at a moment's notice, at any given time.

Trying to determine whether the axe was a weapon or a tool first is a little like the chicken and the egg parable, but one thing is for certain: Any instrument that could chop down a fifty-foot tree found little challenge in chopping down even an armored soldier. There aren't many materials, nor many beasts, that can't be dealt a fatal, finishing blow with a single heartfelt swing of the axe.

Axes are not just axes, even though you can bet that some form of this tool will prove itself to be vitally important to any lumberjacking endeavor. There are a few things that every lumberjack, regardless of how experienced, should know about using axes to chop wood—or anything, for that matter. We kept gray wolves, under license, for eighteen years, showing them as a no-fee public service (we asked only that people pay attention as we lectured them about the truth concerning these smart, social meat-eaters).

The point of this digression is that during this almost two decades of loving our small wild pack (they were so not dogs, and light years from being tame), we fed them red meat every night of their lives. I butchered a lot of animals, and our pack ate a lot of meat that came from hundreds of road-killed deer, scraps from hunters that included bear, deer, moose, lamb, goats, cows, rabbits, and even fowl, like turkeys or geese (some fish-eating birds, like blue herons, cormorants, and mergansers, are inedible by wolves, and most carnivores). In-season, a substantial volume of scraps from the generosity of small businesses, like meat-cutting master Linda Penfold at the Walloon General Store, in Walloon Lake, Michigan, who showed her heart to be at least as big as our wolves.

Carcasses—rib cages, necks, pelvises we just hacked into smaller, more conveniently dividable pieces, using one (or more) of a dozen axes that work for a living around here. "Green" uncooked bones are actually easily digestible by both wolves and dogs, but chopping the femur (or skull—wolves love frozen brains in the winter) or any heavy bone is simply too much to ask all but a few knives or machetes, and there's always a chance of chipping a cutting edge.

Axes are the tools for that job, and, especially for the sometimes 50-pound blocks of just meat and fat scraps that began as small throwaway scraps in a plastic bag, but have now become as hard as Machu Picchu ruins in Peru. For those blocks of clear plastic-encased meat, which must be reduced to as close to mouthful-size morsels as possible to keep animals from exerting too much energy to digest meat that's two to three times colder than the freezer from whence they came, I reach for the 50-pound, hardened steel Dayton-style axe that impacts with eye-popping power, easily splitting a boulder of meat in twain with two to five chops on a twenty-below (Fahrenheit, of course) day, with a wind chill that dropped felt temps 20 degrees below that. A regular 3.5- pound felling axe head might bounce right off meat scraps that solidified, even if it has been made extremely sharp.

The Role of An Axe

The most fundamental and most important tool for reducing even large trees—and almost any other seemingly immovable object—to pieces that can be transported to somewhere else, is one that'll fit in the trunk of the most compact car.

An axe has been clearing the way for humans since before smelted metals existed, and with today's advanced metal alloys and super-strong handle materials, it's more efficient than ever before. Virtually no blockade—whatever it's made from—cannot be broken with an axe. That's why firefighters, who might have to break through brick walls and steel doors, still retain an axe as their most iconic tool of the trade.

But as axes have gotten better, so, ironically, has the number of people who are proficient with them dwindled at a greater rate than typewriter repairmen. That isn't intended to insult anyone, it's just recognition of the truth that very few people today have ever needed to use an axe to cut or split firewood.

The most fundamental concept behind chopping wood is to remove as much of the material as possible with the fewest blows and the least expenditure of energy. Few chores are more bone-wearying than chopping wood, so economy of motion is the order of the day.

Axe Types

There are numerous types of axes used for timber-cutting, most of which are named for the region in which they were first seen; some of them are purpose-built to perform one function, while most are engineered to be as versatile as possible.

In all, there are about forty-two different axe head configurations. Some are built specifically for felling trees: These were typically not very broad, so as to take narrower, easier-to-twist-free, bites out of a tree trunk. Some were broad, generally to take big bites out of smaller trees.

Some were double-bladed, with each side sharpened using a different chamfer, enabling them to be more multi-functional (these were the preferred axes of homesteaders and settlers). Some had rounded corners to make them more easily peel bark without digging too deeply into wood. And some were simply the product of unique personal preferences, custom-made by local blacksmiths to the suit the personal nuances of an individual lumberjack.

Today, less than half a dozen working axe designs are made. Although there do seem to be more than a sufficient number of fantasy battle axes and more or less weird-looking "tomahawk" designs, these aren't of much use for serious lumberjacking. With a few exceptions, they're poorly made, poorly designed, forged of substandard steel, and some look to have been designed by comic book artists. When you by an axe for working, be sure you aren't getting one that was only intended to be admired.

One of the wierder designs in splitting axes, the Lever Axe (R) from Vipurkirves is effective, but its single purpose and unwieldy design limits its popularity.

Except for cast-iron trade axes, which, as the title implies, were mass-produced as fodder for trade with Indian tribes and poor farmers, axes made before the twentieth century were forged individually by blacksmiths. No two were exactly the same, and no two were of precisely the same alloy. Nor could they have been heated-treated or tempered to exactly the same specifications. Some were malleable (soft, hard to break, easily dulled), some were hard (very sharp, but prone to chipping).

Then steel mills made the scene in the first years of the twentieth century, and that changed everything. Literally. Until then, steel was a laborious process that required real blacksmithing skill and was actually a trade in itself. Now, hammer-forging with multi-ton presses enabled mass-production that was crude, mostly impossible before the Industrial Era. Until then, any specific categorization of axes as this or that type was ambiguous at best, because every one of them was essentially custom-made. And by today's tightly controlled quality standards, the best of them wasn't especially good.

Michigan Axe

A Kobalt-brand, Michigan-style axe with a 33.5-inch fiberglass-core haft, and a 3.5-pound head is typical among felling axes today. (Photo courtesy of Lowe's).

The name Michigan derives from the Ojibwa Indian word *meicigama*, meaning great water, a reference to the heavily forested Great Lakes region where it was first made and used.

That the Michigan style of axe has survived to become the axe that you'll most likely find in any hardware store in the nation is an indication of how popular it has been throughout history. Today, Michigan axes are made in single- and double-bit configurations, both typically with a 3.5-pound head. The double-bit design is attributed, in lumberjacking lore, to the giant woodcutter, Paul Bunyan. It was said that the twin cutting edges allowed him to do twice the work.

In reality, the double-bit axe was useful to woodcutters and particularly to pioneers who might be cutting down trees in the day but splitting firewood for the stove at night. One edge was kept razor-sharp to enable it to take maximum-sized bites out of tree trunks. The other was a bit rounded, so that it could perform as a splitting wedge, but not bite so deeply into an upended log being split for firewood that it would become stuck tightly. Lacking a flat side, a double-bit couldn't be freed

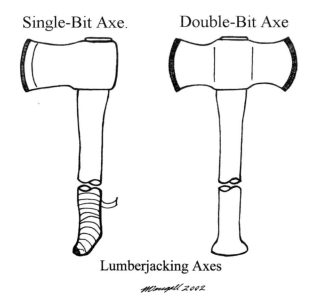

Single-Bit Axe.　　Double-Bit Axe

Lumberjacking Axes

Single-bit and double-bit felling axes.

Michigan-style double-bit axe, modernized with a rust-resistant coating and a fiberglass-core, polyethylene-bonded handle.
(Photo courtesy Ace Hardware.)

by pounding it through, like a splitting wedge. There are tales (usually pretty tall, most of them) of powerful loggers who sank their axes in so deeply no one could ever extricate the bit again—sort of an updated spin on the old Excalibur fable.

The single-bit Michigan axe isn't remarkably unlike other single-edge axes, except that its corners are more rounded; its back side, used for light sledge-hammering duties, like driving a splitting wedge, is more convex than other types. The theory is that a rounded surface allows striking force to be concentrated over a smaller surface area, while at the same time maximizing the area of its cutting edge. It also meant that the person swinging needed skill, because it was easier to strike a glancing blow.

A working lumberjack's axe, like this Michigan single-bit model, is sharp enough to cut meat, so be careful.

In reality, Michigan has become a generic term, used to identify axes in general. I have a large, 5-pound head, single-bit axe whose lines identify it as a Dayton axe, but it was listed in the catalog from which it was ordered as a Michigan axc. It really doesn't matter, because any difference in performance is imaginary, except that its more massive head delivers considerably more power to its target.

Collins Axe

This is the axe that seems most typical in national-chain hardware stores, and it's the cause of some confusion.

Collins is the name of an axe-making company whose roots go back to its start in an old gristmill on the Farmington River in Canton, Connecticut, in 1826. Before Collins began mass production of finished axes, unground steel heads and "trade" heads of cast-iron were imported from Europe, or axes were individually made by local blacksmiths. A good, sharp steel axe was highly prized.

With the success of Collins axes—and machetes, adzes, and most other edged tools of the time—Canton became the thriving community of Collinsville, and the biggest supplier of axes and machetes in the world.

Collins was bought by in 1966, after a long, sometimes arduous business history, by Mann Edge Tool Company. Mann sold the Collins brand in 2004 to Truper Herriamentas, located in Mexico.

Today, Collins axes have almost become synonymous with axes, even though they are, in fact, only one brand. And even though the company manufactures not only Michigan types (double- and single-bit), Crown Pattern double bits, Dayton single-bits, and Pulaski types, with an adze bit opposite a single bit axe, Collins was the first, and for that they deserve historical recognition.

Hybrid Axes

Aside from changing heads from iron to steel, developing the technologies to mass-produce them, and fitting them with fiberglass handles, axes have remained relatively unaltered for a century.

Then, in the first decade of the twentieth century, the Gerber division of Fiskars, a Finnish-based company best known for scissors—in a country that's, historically, best known for producing lumberjacks—decided to modernize the axe.

For centuries, axes remained more or less unchanged, until Gerber (division of Fiskars) redesigned axes from the ground up.

Gerber's Splitting Axe II started the ball rolling with a truly revolutionary design. It has a uniquely wedge-shaped, hammer-forged, stainless-steel (a first) head that's mounted with a band over the top, so that it cannot fly off. The 3.5-pound head's flared-out sides help to protect against the handle damage that's common when attempting to split gnarly, knotty firewood logs. A unique design band over top of the head mounting design negates the possibility of the head flying off, and because there's no hole through the head, it can have the same weight (and chopping power) in a smaller package.

Axes pull multiple duties around here—like chopping this frozen block of venison scraps into morsels that require less energy for the author's licensed gray wolf pack to consume.

Gerber's innovative hollow, 28 ½- to 36-inch, fiber-composite handle is pretty nifty, too. It's hollow, which not only doesn't make it weaker, but actually increases its surface area and, therefore, its strength. At the same time, lighter weight means less inertia at the start of the swing, which translates into greater velocity at the head and substantially increased chopping power. More chopping power, of course, means fewer chops to get the job done—Dan'l Boone wished he'd had this axe.

To finish the package, there's a snap-on molded cover with a carrying handle. This cover is also used on the Splitting Axe's two little brothers, the XL Axe II, a camp axe with a canoe-size 23 ½-inch handle and the Sport Axe II, a 22 ½-inch hatchet.

The Fiskars Super Splitting Axe II is a prime example of how modern technology is keeping pace with mankind's second-oldest tool (next to a rock hammer): Stainless steel, computer-designed head, innovative, super-strong head mounting system, hollow (actually stronger than solid) polycarbonate handle, with a 36-inch stroke make this axe feel like having a super power when splitting the gnarliest wood.

Jack pine is one of the most hated, awful trees on the planet, good for nothing except the manufacture of OSB (Oriented Strand Board), and needing to be cut because only the biggest trees fall -literally and figuratively - victim to the native jack pine budworm. Splitting it is a chore that fights you with every fiber.

The best splitting axe ever devised, according to the author. With its 36-inch "nearly indestructible" (according to the manufacturer) polycarbonate handle and ingeniously-flared stainless steel head, in a no-slip mounting system, McDougall rates this axe almost like having a super power.

Then there's the biggest brother to these, the all-black Fiskars Super Splitting Axe. This has the groundbreaking head design found on Gerber's Splitting Axe II (Gerber is a division of Fiskars), but with a 36-inch polycarbonate handle that appears to be indestructible. Splitting wood with this axe is like having a super power and its extremely sharp stainless-steel head isn't a slouch when it comes to felling, either. Prices range up to about $55. These axes merit individual recognition here because of their revolutionary designs, and because they've performed so darned well for me, personally.

Lumberjacking is stereotypically a man's occupation, but there remain a few environments where women need to be handy with the tools of the trade, too.

Axe Sizes

There are also, generally speaking, two different sizes of axe head: the standard 3.5-pound head and the 5-pound "Timbercruiser" head. Depending entirely on a user's skill, the heavier axe is actually preferred by some, because it hits harder. Harder hits mean more and larger chips with fewer swings—*if* the blade is sharp and *if* the person using it has skill. The heavier head is heavier to carry, and that limits its use by backpackers and woodsmen.

Mentioned elsewhere in this book, beware of the "competition" type axes, with extra-large heads made from alloys that are better suited for skinning knives. These axes aren't made for chopping green wood or chopping through hard knots. They cost many times more than everyday work axes, and the edges are both too brittle and too fragile to chop live trees, cut limbs, and generally take the abuse heaped onto a lumberjacking axe without sustaining damage. Think of it in the same terms as a stock car—in the sort of competition for which it was designed, a racing car blows the doors off a floor-model SUV, but for everyday use, it's not well suited at all.

All things being equal, the longer the handle, the better. Although some fellows might take exception to the term, a popular size for all-around backpacking, camping, and canoeing is an axe that was one called a "Boy's Axe," because it has a shorter handle, sometimes a lighter head, and was a pound or two less to carry.

A Boy's Axe (also, more acceptably called a Canoe Axe) is still popular with outdoorsman. It has considerably more chopping power than a hatchet, but is less to carry. Personally, I believe the increase in power is worth carrying a full-size axe, with as long a handle as possible.

Hatchets - small axes: farthest away is the outstanding Gerber, built for anything you might throw at it; the SOG (middle) is a superb do-everything model, with a sharp recessed into it handle; the TOPS, nearest, is a lightweight tool for light trimming.

Splitting Mauls

Typically, firewood is created by taking a section of log, usually about 18 inches long, and splitting it in half, then splitting the halves again into quarters. Quartered thus, the wood burns hotly and evenly.

The Roughneck™ 8-pound splitting maul is a nice middleweight choice for making firewood from logs. (Photo courtesy of Northern Tool.)

Movies make this process look simple, but those pieces being split by actors on the silver screen were specially selected from sections of seasoned, dried soft woods—like white pine, cottonwood, poplar, etc.—chosen because they have perfectly straight grain that splits as easily as butter. Real firewood tends to be knotty, with grain as crooked as an election-year politician. An actor with a double-bit axe can't split it, no matter how handsome he might be. After one swing, John Q. Lumberjack will probably find his axe head buried deeply in wood and stuck tight.

Ironically, the hardest, most hot-burning and long-lasting firewoods—oak, beech, maple— tend to be the most gnarled and hardest to split. A typical axe isn't enough for these; you need an extra-heavy, wedge-headed splitting maul.

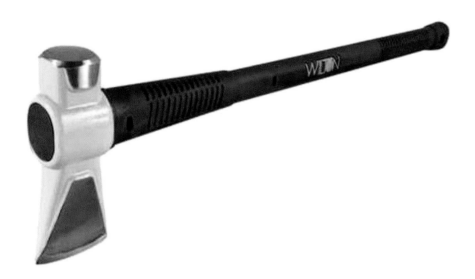

A smallish Wilton 6-pound splitting maul. (Photo courtesy Northern Tool Company.)

To begin, stand the log being split on end atop a splitting block, which is itself just a large diameter log, 1- to 2-feet long (depending on your personal preference, and what works most comfortably for you), placed on end atop the ground. A splitting block ensures that a section being split is sitting on a solid base, and that all of a blow's force is absorbed by the log being split, not dissipated into and muted by the earth below. Elevating wood atop the block also aids splitting by putting it at the power apex, the correct height to receive the most impact from each swing.

When using a splitting maul—or an axe—let the head do all the work. Don't try to power the maul through the wood, or you'll wind up exhausted after just a few swings.

Rest the maul's head on the ground in front of and between your feet, which are spaced about shoulder's width apart. Place your weaker hand at the end of the handle, as close to the end as possible, while still being able to wrap all four fingers around it—just above the flare of the handle's toe if your maul's handle has one. This hand steers the maul as you swing.

The other, stronger hand is your power hand, the fulcrum. This is the hand that lifts the maul's head in preparation for each swing; it should be placed in the middle of the haft—or as close to the middle as is comfortable for you. Closer to the head is okay for longer-armed people, because that makes it easier to lift the tool's head, but never allow your power hand to grasp the handle closer to its end than to its head, as that changes the leverage and makes it too hard to lift its head.

Bring the maul up, directly over your head, or as close to directly as you're naturally comfortable with. Then, in one motion, barely stopping to reverse direction, bring the head back down. As you bring the head down in arc, apply force with the power hand while sliding it downward, toward the guiding hand. Stop applying force just before the maul's head strikes wood, letting the force of inertia do the work. When the maul makes contact, your power hand should be touching your guiding hand and you should only have enough grip on the tool to prevent it from slipping out of your grasp.

Hatchets

Smaller than an axe, a hatchet is more portable and better suited to light cutting duties, like blazing (stripping away bark from) trees, to identify them for cutting, or trimming away branches.

Axe or hatchet, it works most effectively for cutting wood when it's sharp, and a sharp blade needs to be covered when carried afield.

One of the *Standing Orders of Captain Robert Rogers* (circa 1789) of Robert's Rangers fame read, "Let the enemy come till he's almost close enough to touch. Then let him have it and jump out and finish him up with your hatchet."

For an average woodsman, a hatchet has always had more peaceful uses, but even today a hatchet is *the* tool for general utility with most outdoorsmen. It has been said that the lumberjacks and settlers of old wouldn't even make a visit to their outhouse in the middle of the night without first having their trusty hatchet in one hand.

Traditionally, a hatchet was a short-handled single-bit axe with (usually) a smaller, easier-to-control head. Today, hatchets take on numerous forms, from the Drywall Hammer (that became a fad in the '80s, when it was sold under the name "survival hatchet") to the Roofer's Hatchet that is engineered specifically for work with shingles.

The SOG Back Country Axe is actually a light hatchet, but with a small, sharp saw in its handle, it's an ideal companion for light duty tree cutting chores.

Making large pieces of wood into smaller ones is the mark of usefulness for any axe or hatchet.

Machetes

If the branches are less than an inch in diameter, a short, heavy machete might be used. Light machetes, like the GI Jungle Machete, are not good choices for wood. The more expensive of these—i.e., the military issue types—are made from hardened 1095 spring steel that gets sharp and holds its edge well enough to skin a deer. But, while it has proved its mettle against bamboo and in tropical jungle, the high-carbon blades are just too brittle and too light to be effective against wood. Every one of them has snapped off, about 6 inches ahead of the handle.

Softer-steel machetes, like those sold in department stores at bargain prices (usually about $10) tend to bend before they break, but a day of chopping tree limbs leaves their edges so blunted that they won't cut a stick of butter. In addition, these inexpensive machetes are usually fitted with hollow plastic handle slabs that vibrate uncomfortably in your hand upon impact with wood.

Favorites of workers on Christmas tree farms, where cutting off limbs might be an all-day chore, machetes are short enough to enable a swing in tight spaces and weighty enough to allow a powerful chop over a short distance. Favorites include Ontario Knife company's SP-8, KaBar's Cutlass Machete, and TOPS Knives Power Eagle 12.

How to Use an Axe

If you want to see the best ways that an axe should *not* be used, watch any movie which depicts an actor supposedly chopping wood—in the takes that cutting editors choose to show you in their movie, the actor (axe-tor?) strikes precisely the same place on the log he's supposed to be halving, or the tree that he's supposed to be felling.

While it's commendable, even handy, to be capable of hitting the same place on a log over and over at will, that doesn't do much to make one piece of wood into two. And as anyone who has tried it can tell you, swinging an axe is one of the most exhausting jobs that you'll ever try. It behooves a wood cutter to get the job done swiftly, efficiently, and in the least-exhausting a manner as possible.

The Critical V

The trick to cutting through a log, whether vertical or horizontal, is to begin with a wide enough wedge—that is with a wide enough distance between the initial, or *setting*, cuts at either end. The idea is to remove chunks of wood until the log or trunk that's being chopped is in two separate pieces. It follows that the bigger the chunks of wood that you can remove, the faster the chore will go with the least amount of swings.

The width of a beginning cut should equal the diameter of the trunk being halved; that is to say, if a log to be chopped in half is 12 inches in diameter, the distance between the top and bottom (standing) or left and right (lying down) cuts should also be 12 inches. As you cut more deeply into a log, the cut will become naturally wedge-shaped, like a "V." The closer the V gets to the other side, the narrower it naturally becomes and the smaller the chips of wood being removed by your axe. It the apex of the V comes together when you're only halfway through, then you'll just stop removing chips of wood at that point.

If that happens, move your cut back to the outer layer of the tree and simply widen it, taking off wood from either side/end until the V is wider and the vector, where its sides come together, is deeper within the tree.

It's All in the Swing

An old Duke Ellington song says, "It don't mean a thing if it ain't got that swing."

That's a fact when it comes to chopping wood, but one facet of lumberjacking that actors seem incapable if imitating is a proper axe swing. Correct axe technique is essential, because chopping wood is exceedingly hard work, capable of reducing even an athlete to a sweaty and exhausted lump in no time.

The image of a lumberjack possessed of the bulging, venous muscles of a steroid junkie is as unrealistic as Hollywood's portrayal of cops, bikers, and chefs. While it is a fact that you'll never see a working lumberjack who's a couch potato, the secret to wielding an axe isn't strength, but technique. It isn't the man (or woman) who's delivering all that power, but the axe.

To begin, the master, or strongest hand, grasps the axe haft in its center, while the weaker hand grips the end of the handle, just above where it flares outward at its toe or end. Then, the axe is brought back over the stronger arm's shoulder.

When the axe is brought forward to strike, it is powered by the master arm. But that hand slides downward as the axe head describes an arc that will bring it into contact with its target. When the bit bites into wood, directly in front of its wielder, at the apex of its swing, both hands are gripping only the very end of the handle, the master hand atop, and touching, the weaker hand. The toe of the handle keeps the instrument from flying outward, out of your hands.

The axe's power, which can be devastating, comes not from the strength of your arm, but from the velocity of its head, from the angle of the cutting edge as it bites into wood and from the same follow-through that gives a golfer control of a golf club's head at the crucial moment. Don't let off when the axe head strikes, but drive through, as if you were trying to cut all the way through the tree with each swing. The force that gives an axe bit the power to drive deeply enough to remove wood derives not at the point of impact, but during the swing.

Do the Twist

Finally, when you've driven the axe home and the head is buried as far as you can send it into wood, give the end of the handle, where both hands are positioned at the end of the swing, a forceful twist. The purpose of the twist is to wrench free a large chip of wood, if not to break it free completely, then to loosen it so that it will fall out on the next one or two swings.

Generally, the direction of the twist is a downward motion, but if the axe head sticks in place, give its haft a sharp yank upward to free it, and to free or loosen the chip of wood above or below where its head is embedded. Watch any lumberjacking exhibition, where wood chopping contestants remove fist-size chips with every stroke, and you'll see that they give the ends of their axe handles a sharp yank to lever-free a wood chip, which the following swing usually sends flying.

If this lateral twist sounds like it's rough on an axe handle, be assured that it can be, so don't wrench a stuck axe bit back and forth with all of your might—just enough to free it for another chop. Even so, be prepared to replace broken axe handles; an axe that's used shows that it's used, and you aren't chopping vegetables.

A lady only recently remarked to me that many of the axe handles that she'd seen in the tool sheds of "old-timers" looked remarkably like fine-grained furniture. Her implication was that anyone who broke their axe handles must not know how to use an axe. I pointed out to her that, while it's certainly true that novices tend to be overly hard on axe handles, maybe the reason the axe hafts that she spoke of were in such good shape was because they had achieved age without use.

Chopping wood doesn't require extraordinary talent. Anyone can learn to do it—even with just one arm (one of the tricks I use to show off)—but it isn't a job that an average twenty-first century citizen needs to perform, so most people have no experience doing it. Like cutting grain with a scythe, or making baskets from the free-throw line, chopping wood takes practice.

Standing Trees

Before I wrote my book, *The Log Cabin*, I had to first build the cabin. That meant that before it could be constructed, I first had to take the materials for it from the surrounding forest. Standing live trees (poplar, in this case, because they'd grow back quickly and the forest wouldn't miss them) needed to be chopped down, then chopped to length and dragged to the building site. When I was finished, a one-acre area around the site was literally paved with wood chips.

When halving a tree trunk that's vertical (chopping apart a fallen tree needs a slightly different technique), begin with a downward- (inward) angled cut from the top side. Make the cut about two feet off the ground or at a height that seems natural for you: a spot that's natural to hit repeatedly without too much strain on your back muscles—believe me, you'll get plenty of muscle strain without adding more than is necessary.

To repeat the advice given above (because it's important enough to bear repeating), with the axe's edge buried in the wood of the trunk, twist downward forcefully. The space between each initial cut should be equivalent to a trunk's diameter. After the second cut, giving the axe handle a violent twist downward should break free a large chip of wood between the two angled cuts.

Another pair of angled cuts, driven just inside the first two, followed by another strong twist of the axe's handle, breaks free another large chunk of wood. Two more inward-angling cuts, another twist, and so on. This v-shaped initial cut is known as the "felling notch." The purpose of a felling notch is not only to weaken the trunk, so that a tree will topple, but to determine in which direction it will fall.

A felling notch cut into a large tree that is already roped-off, under tension, to a come-along winch.

When the felling notch reaches about two-thirds through the trunk, stop chopping on that side and go around to the opposite side. Begin chopping another wedge into the trunk, about six inches above the apex of the first cut, on the opposite side. This is the felling cut. Chopping the felling just above the V of the first cut is important, because it creates a step that prevents the trunk from slipping backward, toward you, as the tree falls.

In movies, or more recently reality shows, that depict lumberjacks using either hand or power tools to fell trees, it's romantic to show them free-felling a tree, usually to cry of "timberrrr . . ."—the lumberjack's version of "Fore!" But I've always made it a point to never attempt to free-fell a tree that I couldn't lift off its stump. As the story at the beginning of this book illustrates, a falling tree—no matter its size—is dangerous. And even if it's 6 inches in diameter, it can come down with killing force. If the direction of its fall isn't strictly controlled—the techniques of which are explained further on—and if it doesn't start to fall where you'd intended, there isn't a damn thing you can do to stop it.

Making the initial cut for the felling notch; the tree is already tied-off to a winch, and anchored to a strong tree.

Finishing the felling notch.

Limbing Downed Trees

A defining feature of trees is that they have branches and, before a tree trunk can become a log, its branches must be removed. This is a more dangerous job than it might appear. It's even more dangerous not to remove those branches, though. Not only do they pose serious tripping hazards, it's almost impossible to work around them, and they snag on everything if to try to move the trunk.

A lot of experienced (and, unfortunately, inexperienced) wood cutters use the same chainsaws that they used to fell a tree to remove its branches. There's nothing wrong with this practice, so long as you recognize the hazards inherent with doing it. Foremost is that de-limbing with a chainsaw requires that you cut with the nose of its bar. Although it's common for chainsaw sculptors to do that—usually with smaller, more controllable machines—using the end of the bar to cut with is never recommended, because doing so virtually guarantees that it will kick back. The more powerful the saw, the more powerful its kick (and the greater the likelihood of its making violent contact with your leg).[1]

[1] According to the Consumer Products Safety Commission, in one year, there were 28,500 chainsaw injuries, and more than 36 percent of those were to the legs.

Safer—to both yourself, and to your chainsaw—and actually quicker, in regard to getting to individual limbs, is to use an axe. This isn't to say that there are no dangers in chopping off branches rather than chain-sawing them, because hitting oneself just a glancing blow with an axe can have dire consequences.

If possible, straddle the downed trunk with the base of its trunk behind you, its top in front. Direct axe blows to the crotch of the limb where it adjoins the trunk, impacting in a lateral direction with the trunk, as close to the trunk as your aim permits. It's possible to remove the limb almost perfectly flat against the trunk with no stub.

In most cases, you can "pull" your swing, that is, don't haul off and exert a full-power strike, as you would when chopping down a tree. Abbreviated, half-power swings permit you more control with less chance of overstrikes that go through a limb and career wildly behind and beyond the limits of your control. If a limb is too thick, take a wedge or two from it and it'll break off—again, very few things can resist being hit by an axe, and not many times, regardless how strong or heavy they might be.

Attempting to cut a large tree whose center is suspended off the ground at both ends, whether you cut from above or below, will pinch, and trap, a saw's bar and chain; removing a large wedge first can prevent that.

Having first removed a large wedge from the top side (size of the wedge is determined by the girth of a tree, but should equal at least half its diameter), it can be under-bucked in half, without pinching a saw's bar.

Limbing with a hatchet or a machete seems to be safer, because a lighter tool is, indeed, easier to control. But it isn't safe; a missed swing can literally remove a digit or inflict a horrendous wound, so never let your guard down or be lulled into an entirely unmitigated sense of security because the tool is smaller.

Probably the best, safest tool for removing all but the largest branches cleanly and quickly is to use a small fixed-blade hands saw, like the Corona Razor-tooth, $25 to $50, depending on length of blade. Again, don't be fooled into thinking that it's a safe tool. Following is an excerpt from my book, *The Log Cabin*, that illustrates the point pretty well, I think:

Because the gable logs were shorter than the wall logs, becoming shorter yet as they rose upward to the peak, they could be taken from the treetops that were lying all around the cabin. I was sawing through one of these with my left foot on top of the log to help steady it when my boot slid on the rain-slick bark. The slip knocked me off balance, throwing my body toward the log just as I was applying power to the saw's forward cutting stroke. The saw's teeth made a zipping sound as they ran across the inside of my left knee, shredding the ripstop fabric of my trousers and laying the skin beneath open to a width of more than an inch. I could see exposed muscle before bright blood started to flow. It reminded me of what a deer looked like when I was skinning it.

Chopping a Log in Half

Once the tree is felled, its limbs removed, and its trunk bucked to desired lengths, something is usually done with it—depending on its size and the application, it's milled (sawn lengthwise) into boards and planks, used as a wall log on a cabin, planted upright as a pole or fence post . . .

If you're chopping it with an axe, there's a technique for that that every experienced woodsman learns, and it's no coincidence that all of them learn to do the job pretty much the same way—like digging a post hole or hammering a nail. Like almost everything in lumberjacking—and in life—there's no hard-and-fast right way, only methods that get the job done; but there are most definitely wrong ways.

First, "set" your cut—the technique is similar to the one described earlier for felling a tree. With your weakest, guiding hand gripping the axe's handle just above the "toe" at the end that keeps it from slipping from your grasp, and your strongest driving hand wrapped around it's middle, bring the axe's head over your driving hand's shoulder. The position is much the same as if you were getting ready to swing a baseball bat—note that the position is identical to the one used to cut down a standing tree.

The first swing drives the axe bit into the log, perpendicular to its length, but at a slightly inward angle. When removing the embedded axe, give its handle a sharp twist to help break the first chip free.

The second cut, made the same way, angles slightly toward the first from the opposite side. The distance between these cuts should be roughly equal to the diameter of the log being chopped in half. Give the embedded axe a sharp twist of the handle to free it, and to break the chip free.

The second chop should land almost directly between the first two. A twist of the embedded head should break a large chip of wood free, or at least loosen it. Another chop and a twist directly into the initial cut on the oppose side breaks another large chip on that side free.

Continue in this manner, chopping into first one side, then the other, at slightly inward-facing angles, giving the handle a sharp inward (usually) twist after each blow and removing as large a chip as you can with every strike.

As you chop more deeply into the log, it'll become apparent why you needed to start with such a wide cut. The V-shaped notch will become narrower and narrower, the closer your axe gets to the other side.

You don't have to chop downward onto a log for the entire process. In fact, it's more expedient—in terms of removing the most and largest wood chips—to chop from the sides, once to get about halfway through. First one side, then step across the log, and chop the other until the log is held together by only a few fibers of wood and it separates with a final chop.

Brush

The author's backyard.

Taking out brush requires different tools. When the tag alders—a small, softwood tree that grows extremely fast—takes over our little "beach" at the Betsy River, there's no choice but to remove them by chopping them out, by hand, as low to the ground, even below, as you can get. Even so, it's incredibly vigorous root system ensures that the jungle will return in a year or so.

Obviously, a chainsaw is out of the question—cutting at or below ground level would cause bar and chain to grind themselves down to uselessness in minutes.

If the goal is just to trim a trail through overhanging tree branches, or light shrubs, a heavy machete will do the trick. For tree branches, swing downward, striking at the crotch, where it joins the trunk. This is the point at which the branch is easiest to separate from the tree. If the branch doesn't sever, it will break most easily at this junction.

Trimming Branches from Standing Trees

Trimming branches from standing trunks, as opposed to trimming them from a felled tree, is usually a one-hand operation, because an axe is too long and too heavy and you should never, ever cut above the height of your shoulder with a chainsaw.

A hatchet, in fact, does a better job at this task, with its heavier head and more forceful impact. But its narrower cutting/chopping edge is a bit more difficult to aim, it's more tiring to use, and misses are very dangerous; hitting oneself in the thigh is a common injury—and there are few minor injuries with a hatchet.

For this reason, branches should be trimmed using the longer cutting edge of a heavy machete. Again, aim at the junction, where limb meets trunk, with the center of the blade. Do not try to place the point of impact near the end of the blade, as you have less control of a blade there, due to leverage.

For removing shrubs at or below ground level, an axe is the best choice. Its longer handle spares you the back and knee pain of bending over or dropping to a knee. Digging in the dirt will, of course, dull the blade, so be prepared to resharpen it. And watch your toes; lore has it that killer Clyde Barrow sacrificed two of his toes off to get out of work at a prison farm—the truth is that he likely cut it off with a missed swing.

Maintaining an Axe

When canoeists beach their vessels at the end of a day's paddle, they turn the craft bottom-side up. There was once a logical reason for this tradition, even though the practice is no longer necessary with today's metal and synthetic boats. When canoes were made from birch bark laid-up on wooden frames, or hollowed out from whole logs, both of these natural, porous materials soaked up considerable amounts of water; they were turned upside-down at the end of the day, because they needed to dry.

Likewise, the simplest method of protecting an axe's cutting edge is something that woodcutters still do today, often without knowing why: simply burying the cutting edge into the top of a chopping block, or into a log, to protect it from oxidation. Ideally, an axe should be brought out of the weather and stored in a cool, dry place when not in use.

And then there's real life, when you're caught in the rain and the work still has to be done. When the storage shed must be the bed of a pickup or a drafty tool shed. Wooden handles need to be protected from drying, cracking, and shrinking in their heads, and subsequently flying off mid-swing.

Loose Axe Heads

A loose axe head can be extremely dangerous, and it eventually happens to every axe, if the handle doesn't break first. Having a 3.5- pound (or heavier) piece of sharpened steel go flying across your work site with the speed of a baseball pitch is guaranteed to leave a mark on whatever it strikes.

An axe fitted with a wooden handle is subject to weather. Specifically, drying and cracking, with the end result being that the handle shrinks inside the steel head, becoming dangerously loose.

One Swedish company, a maker of mostly small, designer-ish axes, recommends that handles be treated with a linseed-type oil to prevent wood from drying and shrinking. But this process, like some other contemporary maintenance tips, recognizes that most of today's axes don't see the hard, regular use of their predecessors and are, in fact, often ornamental. Linseed oil beautifies a wood handle but does little to improve its real performance.

For a working axe, a major objective is to keep a handle tight inside its head. One of the best good ol' boy treatments is to soak the tool, with the handle mounted in its head, in a bucket of used (or new, if you wish) ordinary motor oil, until the wood is saturated and swollen tightly—this is also a fix, sometimes with the oil heated over a hot plate, for leather-washer handled knives that have become shriveled and gapped. Saturating a handle swells the wood (or leather). It also softens it a bit, but increased tightness from swelling negates that effect. In bygone days (and still today, sometimes), when everyone changed his own engine oil, there was frequently a bucket of used motor oil sitting in the corner of the barn for purposes like that.

Another down-home fix (one of my personal favorites) for loose axe heads is to drill a hole through one side, or both sides, of the axe head, then drive a half-inch-long wedge- or pan-head wood screw, whose head is slightly bigger than the hole, into the wood. The screw's head should fit as flush into the head as possible to keep it from snagging when you use the axe to split wood.

Metal wedges, sold in hardware stores for the purpose of tightening tool handles, can be driven into the top ends of axe, sledgehammer, and other wooden handles to split them and to cause them to fit more tightly in their heads. For lack of proper wedges, two to three large screws twisted into the wooden top usually suffices to cause the handle to fit more tightly, and a few old-timers axes have half a dozen nails driven into their tops to tighten their handles.

I've also had good luck using Gorilla Glue™. This versatile stuff expands as it dries and can actually break some things that are being mended from the pressure it exerts. That swelling characteristic can be used to good advantage if you're trying to get a tool handle to fit tighter. The glue isn't strong enough by itself to fill a wide gap, but when used to coat small wedges of wood, that ore then taped into place, it provides an ultra-tight fit that can outlive the handle it's used on.

Sharpening an Axe

In a lumberjack's world, where cutting tools are used hard and often, honing isn't a recent innovation. It's a fact of life that every generation feels that every discovery made by its members is something profound and new, unknown by previous generations, but re-sharpening skills have been around for more than two million years: since *Homo habilis* used the first stone-edged axe to transform a tree into something that he could use to make life easier.

The usefulness of an axe, or a knife, scissor, or any cutting tool, is seriously impaired if its cutting edge isn't sharp. There was a day, in an era when cutting tools were part of everyday life, when most people knew how to get at least a functional edge on tools. The ability to skin, gut, scale, reap, chop, or slice was no less important than knowing how to make a cell-phone call is today. Now, in a time when few people carry even a pocket knife, folks who can sharpen a cutting edge are as rare as those who can fix a typewriter.

You can prove this to yourself—just run your thumb across (never along, lengthwise, because you'll need stitches if you do that with one of *my* tools) the cutting edge of the axes in anyone's tool shed. Or the knives in their kitchen drawer, for that matter. If I had a dollar for every time I've bet someone that they wouldn't find a sharp edge, I'd be a wealthy man.

The first step is to understand what a cutting edge is, and why it is sharp or dull. The most fundamental explanation is that it consists of two sides that meet at a very pointed apex. The more pointed, then polished that apex becomes, the sharper the cutting edge. It is no more complex than that, regardless of what experts tell you. This is a personal pet peeve of mine, because for decades I've found the less someone understands about sharpening an edge, the more complex they'll try to make someone else think it is.

Applying or restoring a keen edge on anything designed to cut isn't rocket science, but it is a science. There are physical characteristics for any cutting edge that *must* be established before that elusive quality known as sharpness emerges, just as a drinking glass must be watertight or a pool ball must be spherical. Logically, achieving those requisite features is a lot easier if you first know why an edge is keen or dull.

All-Important Bevels

Most important, whatever a blade's design, alloy, or features is the edge bevel on either side of the cutting edge (one side on some Asian grinds). Seen end-on, these two bevels must come together at a pointed apex. The more pointed and polished the V formed by their joining, the sharper the blade.

Applying bevels to a blade isn't as complicated as almost all written instructions make it appear. That isn't intended as a slam to anyone, but merely an observation that most people who know how to sharpen a cutting edge do not know how to explain the process in writing, while most even good writers don't know how to sharpen a tool. My editor at the now defunct *Tactical Knives* magazine, who could do both, lamented that it wasn't really fair for him to evaluate a new sharpening gadget, because he could get a shaving edge from a chunk of concrete.

What he was inferring by that statement is that it doesn't matter how you establish the bevel, or bevels, whether you do it with a file, a honing stone, or even a piece of sandstone, as our ancestors did for centuries. So long as you can abrade away the metal necessary to form a cutting edge that terminates a sharp point, you can apply a cutting edge to anything from a chisel to a butter knife, using any abrasive surface from an industrial-diamond bench stone to a sidewalk slab. Bevels should be more or less evenly applied, meaning that you should remove an even amount of steel, at the same angle, from either side of a blade, but forget precision—first learn the fundamentals.

A tool that should not be used is a bench grinder, and especially by anyone who can't achieve a sharp cutting edge using a manual sharpening "stone." American GIs brought home millions of dollars worth of generations-old Samurai swords as war trophies at the close of WWII, and many of those invaluable pieces of history were destroyed by attempts to resharpen them on a bench grinder. Because it is essentially impossible to control, a belt or wheel type bench grinder should not be used to apply an edge to anything.

Overrated Angles

Humans like absolutes. Parameters, measurements, formulae. By establishing absolute, numerical values for distance, we can know what an inch means in real, repeatable terms that mean the same thing for everyone who uses the measurement. From that, we know feet, yards, miles . . . and we can transfer that knowledge to others, as well as information involving such knowledge.

Lacking other means of transferring the very necessary ability to re-apply a keen edge to cutting instruments, manufacturers, and others who would teach the skill fall back on absolutes. Because a knife, axe head, chisel, or any cutting tool is engineered to exact specifications, re-honing instructions focus on maintaining precise angles when removing the material to restore sharpness.

Too shallow an angle - the honing surface isn't making contact with the edge

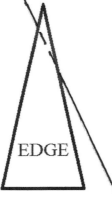

Too steep an angle - you're actually honing away what edge does exist

Perfect Honing angle - surface of the hone drags smoothly over the entire bevel

(c) Len McDougall, 2014

Illustration of a correct honing angle, as well as incorrect honing angles that account for cutting edges that just won't seen to get sharp; regardless whether the tool is an axe, chainsaw, or skinning knife, the principle holds true for any cutting edge. Because there are numerous edge angles on different tools, especially well-used ones, an expert honesman is one who can feel when a honing surface is making full, proper, contact with an edge bevel.

And that is well and good—if you can do it. If you have access to a machine shop or a device that allows you to fixture a blade being sharpened, then apply a precise, matching angle between hone surface and cutting edge, you can do as instructed and grind each bevel to an exact angle.

But if you're in the field—or in most home garages, I promise you that you cannot maintain that level of precision. Try as you might, it's not possible to look at an edge against the surface of a hone, then to maintain a precise angle as you abrade the entire length of a blade. You'd need micrometer eyeballs, a vise-solid grip with both hands, and fixture-like steadiness.

In a nutshell, it's impossible to freehand sharpen by visually keeping the blade—any blade— at a precise angle while you abrade its cutting edge against a honing surface. And it doesn't matter, anyway, because every knife, axe, or lawnmower blade, if it's kept sharp at all, winds up having a different edge angle than the one it was manufactured with—that's simply a function of resharpening, time, and normal wear.

The bottom line, whether we're talking tomahawks, hedge shears, or wire-cutters, is that angles are unimportant enough to disregard them.

The *apex*, however, the point where bevels meet to form a point, when viewed end-on, are of critical importance. No apex, no cutting edge, regardless of the honing angle.

F-E-E-L the Edge

What is possible is to *feel* if a bevel is flat, at the correct angle, to a honing surface. Every stroke is perfectly set against a honing tool, at whatever angle a cutting edge happens to be, because a honesman can feel when the angle is correct. A skilled—and by that I mean experienced, because that is the only way to develop this skill—honesman can sharpen an axe or knife in pitch darkness. He *feels* the edge against his hone.

A dull blade may not exhibit this drag if it has been eroded by weather, poorly sharpened, or perhaps never sharpened at all (probably most hardware store axes I've seen are sold having never been sharpened at all). This means that you'll have to re-establish the cutting edge bevels with a hone. It doesn't matter if you use a file, an aluminum-oxide hone, or even a chunk of concrete, so long as it can abade-off enough steel to apply even a rough, but even, bevel to both sides of the axe-head. You can tell when the bevels are there. Because then a honing surface will exhibit an even, smooth drag as you hone.

Honing Techniques

Grinding away steel, the most fundamental definition of re-sharpening a dulled cutting edge, with a hand hone can be a lot of work. A lightly dulled edge requires less effort to re-sharpen, only because it requires a lesser amount of steel be removed. Heavily used, and therefore dulled tools, like machetes and axes, may need to steel "hogged" off, to use an old machinist's term, even though manual sharpening remains the most preferred method, because it enables you to re-sharpen a cutting tool even in the deep woods.

Files

Mill files are the best tools to use when fast removal of large amounts of metal is required. For instance, a 10-inch mill file is the common choice for sharpening lawnmower blades, shovels, large garden tools, and axes. An 8-inch mill file gives a slightly smoother surface, is better suited for smaller axes, machetes, and hatchets.

Mill files have different cuts or coarsenesses: coarsest is the bastard cut; next is called a second cut; finest is the smooth cut. In addition to its cut rating, coarseness is also determined by a file's size. With the same cut rating, the longer a file, the coarser it becomes: A 10-inch Bastard Cut mill file is coarser than an 8-inch Bastard Cut mill file.

If an axe has never been sharpened, you'll need to remove a lot of steel to apply the initial bevels. For this, I prefer a 10-inch Bastard Cut file, preferably with a handle affixed to its rat-tail to make it easier to use. Fixture the axe—in a workbench vise is best, but just laying it on a solid surface so that the edge hangs over—then placing a foot or a knee on it to hold it in place as you file is enough.

A file should be pressed against a blade in a single direction, from outer (flat) end to inner (handle) end. It should not be sawed in both directions, although some skilled honesmen make it appear that they're doing just that; in fact, they're only pressing against the blade on the outward stroke and merely dragging the file back under no pressure on the return stroke.

When you've applied an even cutting bevel to either side, the most obvious indicator of success will be a keen point that catches against skin when you run a thumb lightly across (never lengthwise, or with) the edge. Visually, a good bevel is on an even, flat surface that runs continuously from one end of a cutting edge to the other. An uneven, or incomplete, bevel is identified by what appears to be a line in the midst of a bevel; this line tells you that a bevel is ground, or filed, at two or more angles. To achieve maximum sharpness, there must be a single bevel.

In general, a shallow, narrow bevel, which is easier to form because less metal must be removed, cannot be made as sharp as a steeper, wider one. A shallow bevel is tougher, because it keeps more metal on the blade and is better for hard use, like splitting wood. It also loses its sharpness more quickly than a broader, steeper bevel, largely because there is less of a cutting edge to begin with.

Sometimes axiomatic the old benchmark of being "sharp enough to catch on a thumbnail" isn't sufficient. A guy who skins a lot of deer, or, especially, pigs (pig hide dulls everything), or a whittler of any medium, needs an edge that'll shave hair of your arm, and one that will keep a functional cutting edge through as much use as possible before it needs to meet with a honing stone again.

A lumberjack whose cutting medium is wood needs those same properties in his cutting tools, to a maximum degree. Sharper bevels, which result in a sharper, more long-lasting edge, take a deeper bite when chopping through wood, even though it's not as strong. These truths are a product of physical laws. Beyond that, there are other factors, like the alloy, hardness, and grain structure of steel, but the steepness of the angle where two edge bevels meet to form a cutting edge is the major characteristic that determines how sharp an edge can be made.

Individual lumberjacks establish their own happy mediums between the two extremes—another reason why sharpening formulae that dictate precise honing angles has little real importance. It's virtually inevitable that you'll change those angles after repeated re-sharpening, anyway; that's why it's so important that you develop a *feel* for when a bevel is lying flat against a honing surface.

Re-Sharpening

A previously sharp edge already has the necessary bevels, but the apex where they meet is no longer pointed: It has been blunted by friction and wear—in a word, *dulled*. In the phraseology of old timers, the apex where the bevels meet needs to be "stood up," brought back to a sharp point by removing just enough metal from the angles to restore it to its original keenness.

This is a much easier task than setting an edge on an unsharpened blade. You'll be able to accomplish this with just a whetstone. Old-timers knew this process as "standing up" an edge. That, in essence, is just what resharpening is: taking an edge that has been worn blunt—rounded, not pointed, anymore—or peened to one side, usually by impacting a blade against a hard surface, and bringing its sides back a keen apex.

Aluminum Oxide Hones

The favorite tool of mine, and most tool sharpeners—even for their finest skinning and hunting knives—is a double-sided aluminum-oxide stone, sometimes called a "carborundum" stone. Not really a stone, in the sense that the legendary Arkansas Oil Stone (hard, used primarily for polishing very sharp cutting edges), an aluminum-oxide stone is made by compressing abrasive granules, then baking them under pressure into different shapes. For that matter, what is called aluminum-oxide might not be aluminum oxide.

For hand hones that are used afield, the usual configuration is a circular, puck-shaped, handheld model, about 8 inches in diameter. These retail for under $10 and are favorites of backpackers and backwoods lumberjacks for whom portability may be critical. Most popular for re-honing axes and hatchets, you're likely to find one of these do-it-all tools in every lumberjack's axe-sharpening kit.

The second form of hand-hone is a "bench-stone," a rectangular tool of varying widths and lengths, but usually 6 to 12 inches long, by at least 2 inches wide. It's known as a bench stone because most people find it easier to lay it flat on a workbench when they hone rather than holding it in a hand. These, too, are inexpensive, ranging from about $10 to $15.

Most abrasive hones are double-sided, with coarse, 150-grit, and fine, 240-grit, surfaces that enable you to get a razor-sharp edge from even blunt, hard-used axes and machetes. The coarse side is used to grind a rough edge onto a blade, then the finer side is used to polish it to keenness. The smoother and more polished the bevels that form an apex, the sharper a cutting edge.

Most skilled honesmen usually apply hone to blade in a rotating motion, grinding one against the other. Again, it doesn't matter how you abrade away the metal necessary to form bevels that meet at a keen point. Grinding in a circular motion is merely more efficacious, covering more ground so to speak. If the blade being sharpened is small enough—a 6-inch hunting knife blade, for example—the blade is rotated against the stone. If the blade is large, like an axe or machete bit, then it's usually easier for most folks to grind stone against blade.

It's easier to keep a bevel flat if your circular grinding motion is *against* a cutting edge, meaning that the sharp outer edge is pushed into the hone, as if you were trying to shave a layer from its surface. This is usually the accepted method for sharpening a knife.

Grinding a knife *with*, that is, away from the edge, makes it likely that you'll end up with a *cantled*, or convex edge. A cantled edge rounds outward as it comes to a point and is actually preferred on hard-working tools—like axes and machetes—because it leaves more steel on a blade and is therefore stronger. All things being equal, it cannot be made quite as sharp, nor will it maintain a really keen edge for as long, but it's the best edge for hammering tools that take a pounding in regular use because it's strong and it holds a working edge throughout a lot of use.

Stropping with the edge is also done against a (usually) leather belt when a sharpened knife is being polished to shaving sharpness. A loop of rope or a strap affixes the leather strop to one foot, while its opposite end is held against the inside of your opposite knee. Then the cutting edge of the tool, whether it's an axe or a skinning knife, is scraped smoothly against the strop "stand-up" the apex where the bevels form a point.

Grinding an edge with a circular motion—against or with an edge—demands some skill if you're to maintain a consistent angle and achieve a sharp edge. An easier way to apply an edge bevel, especially to a very dull blade, where a lot of metal must be taken off—like trying to remove a chip in the blade—is simply to grind blade against stone (or vice-versa) with a firm back-and-forth motion. Try to keep the blade at a consistent angle as you abrade it first against its cutting edge, then back, away from its edge. Do this up and down the length of a blade until a consistent bevel, identifiable as an even, unbroken, shiny border, extends from one end of a blade to the other.

You'll be able to feel an edge as it forms by running the ball of a thumb across it—again, never lengthwise, as if trying to slice into your thumb; because if a blade is sharp, it *will* slice into your thumb. A sharp, even edge will catch against the skin slightly. An edge that is essentially sharp, but *peened* to one side, will feel sharper on one side than the other. The solution is to hone the other edge until they're evenly sharp on both sides.

Diamond Hones

A diamond hone is another type of abrasive honing tool. Not usually so coarse as an aluminum oxide stone, diamond hones are more suited to evening a rough edge, and a skilled honesman can get a shaving edge from it.

A product of the latter twentieth century, a diamond hone is essentially just a milled steel plate, to which a layer of industrial-grade diamond dust has been bonded. Evenly spaced holes on the steel plate collect powdered metal from a honed blade, and extend a hone's service life between cleanings.

Like any hone, the best honing and cleaning solution for diamond hones is plain water, with a little liquid soap.

For doing it all—and especially for the big jobs that require a lot of metal to be removed—I still prefer an old-fashioned aluminum-oxide stone. But for quick re-sharpening tasks, a diamond hone is just the ticket. Diamonds are, indeed, forever, and a diamond plate never wears out. If it loses its abrasive quality, just scrub it clean with a little soap and water and it's good to go. I did have a portion of the diamond coating flake off from a Lansky® diamond rod once, but that has been my only cause for complaint in the past decade.

Depending in size and shape, diamond hones retail at about $10, to start. I'd stick with name brands, like Smith's and DMT.

Carbide Sharpeners

A carbide pull-through sharpener is essentially a hand-drawn end mill; it uses two (or more) sharpened carbide cutting blades, configured as a V, to scrape metal away from either side of a blade and transform a dull U-shape back to a keen V-shape. The general design has been around for decades, usually in the form of several interlocking hardened steel wheels whose radii intersected one another at the point where a blade was drawn through them. Today's pull-through sharpeners are much advanced over those old models.

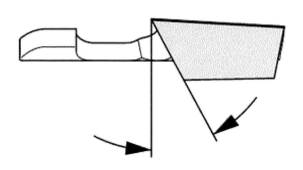

Chainsaw sharpening angle. (Courtesy of Stihl).

Drawbacks to a carbide sharpener include that it has a fixed angle that forces you to adopt just one setting. Many old sourdoughs can be counted on to change the sharpening angle of every axe or knife, right out of the box, because (usually) they prefer a steeper, sharper angle for skinning and other jobs, and edge angles are bound to change as a matter of course as the result of numerous re-sharpenings. Advantages include that this type of sharpener bevels both sides of a cutting edge at once, halving the amount of work and time invested to get it back to keen.

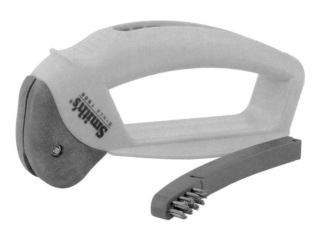

Smith's carbide pull-through Axe and Machete Sharpener is a quick and easy way to apply a very sharp edge to axes and hatchets. (Photo courtesy of Smith's Abrasives.)

Yet, even their fixed angle permits some error and it is possible to get uneven bevels with one wider than the other. That happens when a blade is canted from one side to the other. Always keep a blade as perpendicular, as straight up and down, as possible to a carbide sharpener, and try not to angle a blade as you pull it through the carbide cutters.

Another common complaint is, "It ain't doin' nothin'." As with an abrasive hone, an edge has to first be set. An uneven, rusted, or neglected edge will chatter or slip over the carbides until it has been honed, effectively planed to a level surface that makes smooth, even contact with the hones on either side. Probably the most common mistake with a pull-through (or any) sharpener is giving up before an edge has been set.

Do not press hard against the hone. Set the near end—the end closest to you—of a blade being sharpened into the carbide slot, straight up and down, at 90 degrees to the base. Draw the blade backward. It will chatter, slip, and not feel like it's doing anything at first, but don't quit.

When you reach the forward end of the blade, lift it out of the sharpener and repeat the stroke, starting at the heel (rear) of the blade and dragging it through toward its front. Again, do not press downward any more than is necessary to maintain firm contact between blade and sharpener.

People like numbers and precise instructions, and sharpener manufacturers tend to give customers what they want. But when a manufacturer tells you that it only takes a dozen or so strokes to re-establish a cutting edge, the company means "under perfect conditions, with new, unused cutting or abrasive surfaces." In reality, you'll need to continue to draw blade through carbides (they do get dull) until you feel it bite into steel, until a blade makes an entire cycle, dragging evenly the whole way and making a steady scraping sound. No chattering or slipping, just a steady, even resistance throughout the cycle. This might require more than a hundred strokes, but it will happen, and the result will likely be an edge that'll shave whiskers off your face.

Honing Solutions

Aluminum oxide stones are often used with oil or some other commercial honing solution. A dry stone—providing it is of the grit needed to remove as much metal as is needed—offers all the abrasion required to create a sharp edge. Problem is, stones "load up" with particles of themselves and with metal ground from a blade. When that happens—indicated by a dark, shiny area on the surface of your stone—a stone loses its ability to cut, and it no longer abrades the blade.

A honing liquid helps to keep dislodged particles in solution, which aids in keeping the abrasive qualities of a stone working longer. It also helps to keep a stone clean and abrading more efficiently, because particles that are in suspension can be wiped away with a cloth.

Some authorities suggest specific honing oils and other solutions for use with stones made from aluminum oxide, carbonized silicon, or diamond, but it has been my experience that the best honing fluid for these man-made abrasive stones is ordinary water, sometimes with a little liquid soap to further aid in keeping particles in suspension. Oil tends to work into and glaze the surfaces of these hones, making them more slippery than abrasive, and a glaze is very difficult to remove. Wash your stone frequently to keep it cutting effectively, but I don't recommend using any honing solution but water and a little soap for man-made stones, including those plated with industrial-grade diamonds.

For axes and most wood-cutting tools, you need only a working edge—that is to say; it need not be polished to razor sharpness. If extra keenness should be necessary, you'll want to use a very hard natural hone, like the famed Arkansas Oilstone, to further polish an edge. As its name suggests, this type of hone can be used with oil; its small pores and hardness prevents it from becoming glazed quickly and it cleans easily with soap and water.

Sharpening Systems

A new generation of tools resolves the problem of maintaining proper angle by fixturing, or otherwise controlling, the bevels applied to a blade so they cannot deviate and must meet at a sharp apex. Most system-type sharpening gadgets are comprised of numerous parts and are just too complicated or too cumbersome to be carried afield, especially when hard-used tools might need to be sharpened several times to complete a job. Some, like the Wicked Edge System, about $250, enable even a beginner to get a downright scary edge on any cutting tool the can be fixtured in its jaws. But at this point, the system is just too bulky for field use. Better to learn the manual skills and manual tools explained above.

HAND SAWS AND OTHER HAND TOOLS

According to Pliny the Elder (Gaius Plinius Secundus) in his book, *The Natural History*, c. 77 AD, Daedalus, the mythological Greek engineer who tried to fly with his son, Icarus, was the inventor of the wood saw—and the axe. He crafted the first saw from the jawbone of a snake, and it worked so well that it inspired him to create saws from metal.

That, of course, is mythology, but it does show that the saw is a very old tool—the earliest examples date back to 4900 BC—and that lumberjacking is one of the world's oldest professions.

Buck Saws

Arguably the earliest real saw, as we know the tool, a buck saw gave rise to the term "bucking" logs, meaning to cut whole logs into sections. It uses leverage applied by using a top horizontal member to pull two vertical handles at either end toward themselves, with another beam center between them, acting as a fulcrum. By pulling the vertical members together at their tops, their lower ends on the other side of the center fulcrum are pulled apart.

Underbucking small trees, like this birch, is quick, with no chance of upward kickback, or pinching the bar. But be warned that if a tree is more than 10 inches in diameter, it may pinch, and trap, the bar; to prevent this remove a wedge from the trunk, as described here.

A saw blade affixed to the lower ends of the vertical members is pulled more tightly as the upper member tightens. Kept under tension (much like a modern hacksaw), the saw blade cuts easily through wood without any undue flexing or binding.

Because a buck saw can be operated by one person (or two), it was a vital piece of equipment for the earliest homesteaders. They used it not only to cut building materials whose flat ends allowed a more precise fit than if they were chopped, but for "bucking" firewood with less work and waste than if they, too, had to be chopped to length with an axe. Even so, it was once an axiom that every length of fire-wood be heated twice: once during the bucking, and once during the burning.

Traditional buck saws are still made and used today, but they've been largely replaced by the Bow Saw.

Underbucking small trees, like this birch, is quick, with no chance of upward kickback, or pinching the bar. But be warned that if a tree is more than 10 inches in diameter, it may pinch, and trap, the bar; to prevent this remove a wedge from the trunk, as described here.

Two-Man Crosscut Saw

This is the saw of lumberjack fame, the predecessor to the chainsaw. A long blade, usually 4, 6, or 8 feet in length, with a straight, flat back and an outwardly curved cutting edge, lined with coarse saw teeth, a two-person crosscut saw is fitted with handles at either end, so that if you're lucky enough to have help, whatever cutting job that faces you, it isn't so daunting. There are about 6 inches of fine teeth just below each handle to keep the saw from catching and binding, with more aggressive, coarser teeth around the bow toward the middle of the blade.

A two-person crosscut is intended to take down large-diameter trees, where a long blade is required to reach from one side to the other. A saw blade greater than 3 feet in length is largely impossible for one man to operate, because it must necessarily be thin (the serrations on the spine of a survival-type knife were never designed to cut through wood—the blade is too thick) and it will flex and hang up in the wood as it's pushed forward on the return stroke. With this design, there is a second person to pull the saw back after it travels its full length toward the first person. In this manner, the two-person crosscut removes wood in both directions. Prices start at around $170.

A 3-foot, one-person version of the timber crosscut saw exists as well. Looking much like a larger carpenter's crosscut saw but with much more aggressive teeth, it can be used by a single person. Holes at either end of the blade enable two-person handles to be mounted and the saw to be used by two people. Prices begin at about $170.

Bow Saws

Although I can't recall a time in my life when I was without one of these saws (there are two hanging in my barn right now), I'm not shy about telling anyone how much I hate the bow saw. A modernized version of a Buck Saw, it is also known variously as a Buck, Finn, or Swede saw. Its D-shape frame design enables it to tackle large diameter logs, while coarse teeth—much like the above Crosscut saws—rip aggressively through wood . . . at least, in theory.

The problem is that a bow saw's blade is narrow, it flexes, and it never seems to be taut enough between the ends of its frame. It binds and snags in whatever its cutting, and it has a real tendency to veer to one side as it cuts, making it difficult to buck a flat end.

It helps to keep a bow saw's blade sharpened, taut, polished, and well oiled. Beyond that, it demands skill to learn to cut well with it. As with all hand saws, it doesn't cut on the forward stroke, but rather does so when it's being pulled back. Even so, the forward push causes the blade to flex, and bind in the wood. You have to learn to push it forward with a slight lift that allows it to ride over wood without cutting much, then draw the saw back while pressing lightly downward, ripping through the wood.

Fortunately, a bow saw can be worked by two people as well. This enables it to cut much more efficiently, without so much effort, and with fewer hang-ups.

Razor-Tooth™ Type Saws

This handsaw made its debut from Corona Tools in the early 1990s. Its laser-cut teeth enabled it to have a downright vicious cutting design that rips through wood, and it was adopted enthusiastically by everyone who needed to cut wood. Before long, the three-sided tooth design became a standard for every saw, from tree-pruning types to carpenters' crosscut saws.

These saws are virtually impossible to sharpen, but that isn't really a problem. I used the same Corona 18-inch Pruning saw hard for ten years, until it was finally stolen.

The three-sided tooth saw is now the standard, and it's available in almost every configuration, from pole-type overhead pruners to folding backpack and carpentry saws. Prices begin at around $12.

The author's two favorite axes: (top) a Corona tools Michigan-style felling axe, and a Fiskars Super Splitting Axe II.

This felling axe from Corona Tools has a Michigan-style head, and a tough fiberglass-core, polyethylene handle that's 36 inches long to give it tremendous chopping power.

Cant Hook

Rarely seen these days, and even more seldom used, a cant hook or "cant dog" was once as common to lumber camps as an axe. Origin of this tool is lost in history, but it was almost certainly a case necessity breeding creativity.

Probably the simplest description of a cant hook is a long, stout wooden lever with a hinged iron hook (a "swing bale") at its lower end. By placing the bottom end against a log, the hook digs into the log's side, ever more tightly as the lever is pulled back, toward the wielder. With his

A modernized Woodchuck dual Peavey Cant hook, with two log spurs and gripper-serrations.

This century-old cant hook, a veteran of northern lumber camps, is essentially unchanged from the versions being manufactured today.

muscle power and weight amplified by leverage, a lumberjack could roll or "cant" even a large and heavy log.

A Peavey hook is a modification of the cant hook, created in 1858 by a blacksmith named Joseph Peavey, after he watched "river drivers" riding a raft of logs down the Penobscot River in Maine. It was the river driver's job to grab individual logs with his cant hook and move them about as necessary to keep the raft flowing smoothly with the current, to prevent log-jams. It was exciting enough work (that's where the "sport" of log-rolling originated) without having a cant hook that often slipped before it grabbed.

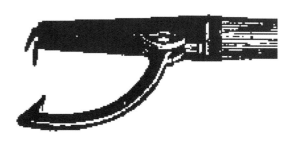

Peavey's simple modification was to add a spike to the lower end of the long handle, like the trash-picking spike used today to spear litter at the sides of highways, only considerably heavier. By first stabbing the spike into the top of a log, it anchored the handle while the cant dog dug securely into the log's side. Before long, cant hooks were being replaced by more efficient Peavey hooks.

A third type of cant hook is the Log Jack cant hook. It differs only in that it has a T-shaped leg attached, perpendicular to its handle, opposite the hook. Its advantage is that it can not only grab a log and roll it, but can also lift one end of even very heavy logs off the ground sufficiently to tie a rope around them.

Cant hooks and log jack hooks are still manufactured and used today, and are available for about $30 to start.

Pickaroon

A Pickaroon (or Hookaroon) is simply a long, stout pole with a usually curved, sharp hook attached to its working end, perpendicular to the pole. A pickaroon isn't made to handle large logs, but rather for snagging, say, firewood lengths from the bed of a pickup with bending or contorting.

There's sometimes confusion over which tool is better suited for moving logs, but the distinction between uses is as simple as remembering that a cant hook can't hook small-diameter logs; it's for the heavyweights. A pickaroon can hook anything, but it can't handle big, heavy logs.
Pickaroons are made in various configurations, and start at about $20 each.

Log Tongs

Log togs are similar to the perhaps more recognizable tongs long used by railroad workers to lift and to carry a heavy steel rail between two men. Essentially two hooks, joined together in their centers by a rivet-hinge, with a carrying handle on either hook's upper end, a tong can grab and lift very heavy objects. Its secret is the scissor-action that pinches harder, the heavier the weight being lifted. If you can lift it, you can carry it without fear of losing your grip.

Originally, tongs were intended to be used by a pair of strong men, but today there are several makes of one-person tongs manufactured that allow a single person to lift and carry logs up to 18 inches in diameter without help. Again, if you can lift it, you can carry it.

One- or two-person loge tongs retail for about $30, and up.

Ropes

Without rope, there would be no logging, because an ability to bind, restrain, attach, and haul is fundamental to almost every building endeavor.

Rope is another commodity that modern people take for granted, but the super-strong, easily tied (and easily untied—that's important, too) weatherproof lines of today weren't possible, even at the start of WWII. The loss of the Philippine Islands was a serious blow to the US Navy, because "Manila" (hemp) rope was vital to ship operations. For millennia, rope had been woven from natural plant fibers (not always hemp) by increasingly sophisticated machinery, and by hand before that. None of these even compared to the toughness, strength, and other positive attributes of the modern polyester ropes that became available to lumberjacks in the latter half of the twentieth century.

Today's version of the log-haulers.

The big boys, pro-foresters and loggers who haul tandem semi-truck loads of commercially cut timber past my house almost daily, rely more on logging chains, large-diameter cables, and chain binders to secure the loads they haul. But it isn't likely that a homeowner, or even a commercial arborist, will need such monstrous equipment or—perhaps more importantly—that they'll have the heavy motorized equipment to make such heavy fasteners work for them. And if you do have a logging job that's too much for the equipment described here (not conceivable for DIY lumberjacking), that's a job for the guys with Log Skidders, Bunchers, and Harvesters (giant, usually tracked vehicles employed for commercial timber harvesting).

Determining the Proper Rope

In 2002, I went into the wilderness and built a log cabin, both for a book I subsequently wrote of that title, and because I sincerely wanted to know what my trapper forebears must have gone through. I kept as close as I could to performing the (as it turned out) monumental task as authentically as possible, using only an axe and hand saws, but reality intruded in several areas.

Rope was one instance where the New Millennium infringed on fantasy. For dragging the 3-ton wall logs I chopped from 40-foot trees, and for hoisting prohibitively heavy roof members in place, my "mule" was a hand winch, with half an inch of *Forestry Pro* rope for handling the loads.

Not remarkably different than the heavy ropes that rock climbers bet their lives on, my rope was12-strand polyester tree climbing/rigging line. A heavy braided jacket—very strong by itself, protected the rope from abrading in the harsh environment and brutal working conditions, while the inner cords lent strength enough to handle working loads of 7,194 pounds, or 32 Kn (kilo-Newtons).

Half-inch forestry rope, rated to work with more than 7,000 pounds, is vital to tree-felling operations.

Knots

Basic Lumberjacking Knots

Square Knot: Fastens two ends together.

Bowline: Non-tightening loop.

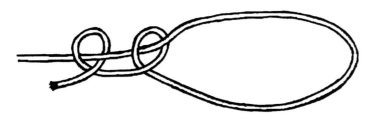

Double Half-Hitch Slipknot:
Noose tightens when pulled.

How many times have you witnessed a load that needed to be secured come untied? If you're an average person, you've seen it numerous times, with everything from moving furniture to tying down lawn equipment. If you live in timber country, you've probably seen it, too, except that these loads are very dangerous to life and limb when they pull free. Tying knots properly isn't just important to a wood-cutter, it's vital.

Deepwater sailors make an art of tying ropes, enough knots to justify entire volumes on the subject. By all means study knot-tying as deeply as you wish, but begin by committing the *Essential Five* lumberjacking knots to memory.

The square knot, bowline, and double half-hitch slipknot.

Square Knot

This is the first knot that anyone who can tie their own shoes learns. With the exception that shoes are tied with a quick-release version of this knot (explained in a moment), every shoe with laces is fastened with a square knot. A square knot is easy to tie, untie, and not likely to come undone when you don't want it to. In fact, a square knot, sans quick-release loops (like most shoes), is the knot I use to secure my working boots, even when temperatures exceed negative 30 degrees and my laces are coated with ice.

Forget about rabbits and holes when learning to tie this most basic knot. Use the diagram presented here as a guide (it also helps to use dissimilar, or different-colored ropes) when learning to fasten a pair of ropes together. Practice it until you can tie it in the dark—really not hard to do—and cannot accidentally make a wrong bend that turns it into the dreaded "granny knot" that can

A square knot is the quintessential knot for quickly and securely fastening two ends of a rope together.

pull tight under load, becoming almost impossible to untie, while, at the same time, coming untied at the most inopportune times.

Begin with two ends. Cross them, the right end over the left, closest to your body. Bend the end closest to you completely around the opposite end, until it is in the same position on the left, closest to your body.

Then, take the end, which is now on your right, and cross it behind the left end. Bend that (right) end toward yourself, completely around the left end until it emerges on the other, far left side.

Pull the two ends tightly and you've tied a square knot. It can be easily untied, even with one hand, by grasping one of the wraps between a forefinger and thumb, then working it loose with a back-and-forth motion.

Slipknot

The simplest, most used slipknot—not just for logging, but for making survival-type snares, for kayaking, and for all-around everyday use where a sliding noose is needed—is the double half-hitch slipknot.

This knot is deucedly simple to tie; many children teach it to themselves. Hold the end of a length of rope in both hands, the bitter end (a Naval term) in one hand, the other hand spaced a foot or so up its length.

Cross the end over the rope so that it forms a loop. Wrap the end completely around the rope through that loop so that about 6 inches of the end sticks straight up.

Take the end and cross it over the rope again so that another loop is formed below the cross. Insert the end through that loop and pull the end until it tightens. The first loop at the bottom is

The ability to use wood products from the environment enabled humans to build houses, traps, machines, and civilization.

now a noose that can be enlarged by pulling outward against either side of the loop, or constricted around whatever it's placed over by pulling against the free end of the rope.

Bowline

Sometimes you don't want a loop to tighten, but rather remain the same diameter when stress is placed on it. For example, when you're tying a foot loop for someone to step into and be hoisted up. A bowline is especially useful when a rope is wrapped around a tall, immobile object, with a loop that will not tighten, but will slide freely all around that object's circumference. The kind of flexible belay that sailors want when tying up to piling to ride out stormy weather.

A bowline knot begins about 3 feet up its length from a free end. Bend a loop at this point, forming a loop. Place the end of the rope through that loop. It should extend through the loop at least 6 inches, but bear in mind that you are now determining the fixed diameter of the main loop below.

Next, with the end extended through the smaller, upper loop, wrap the free end around the junction of the place where the rope crosses itself and through that loop again. Pull the free end and the loop taut against each other to tighten the knot. Voila! You have non-slip loop that will not tighten against whatever it encircles.

Overhand Safety Knot

An Overhand Safety Knot is preferred by firemen and other public safety professionals. It's simpler to make, arguably faster to tie, and it works at least as well a bowline. Best of all, it may be tied when you have access to just a few feet of rope at one end; it cannot be tied around an object.

Tie an overhand safety knot once, and you'll understand why it has been adopted by so many public-safety organizations. Loop the end of a rope around on itself until it fashions a U, a teardrop shape, with 2 feet of the end parallel with the remainder.

Bend the lower length of that doubled U up until it crosses the paired upper end, forming a loop. Stick the doubled closed end through that loop and pull it taut, from the bottom and from the top. You'll be left with a large non-tightening loop with a heavy knot at its top—ideal for placing a foot into while someone pulls you to safety.

Timber Hitch

Don't let this knot's place in the lineup fool you into underestimating its importance; using a timber hitch, sometimes referred to as a choker hitch (or just a choker), is imperative for all logging operations. In fact, it may well prove to be the most valuable knot you know for everything that involves tying with rope.

That's a big claim, but a choker hitch makes it possible to drag a long pole through rough terrain. If you've ever tried this before, you've almost certainly been frustrated by even the tightest slipknot's tendency to simply slide over the length of the pole until it comes free. Considering that I had to drag (with a come-along winch) every 3- to 4- ton log that went into the construction of my log cabin through swamp, leaving furrows as they plowed through, you know that I faced this problem many times.

A timber hitch is essentially a knot amplifier. It uses a small constricting force to hold a more powerful constricting force. The harder you pull against it, the tighter it becomes, until friction multiplies to the point where you can pull even a smooth-sided steel pipe lengthwise along the ground without it pulling free.

Begin with a slip-knot noose placed around the log to be hauled. But the noose isn't enough to do the job by itself. A half-hitch, made by wrapping the long, pulling end of the rope around the log, then inserted back through itself where it meets, multiplies the holding power of the noose, usually to the tensile strength of the rope being used. I've never needed a second hitch, but some people add one, just to be dead sure that there can't be a slip.

Timber-Hitch (Choker) Log-Skidding Knot

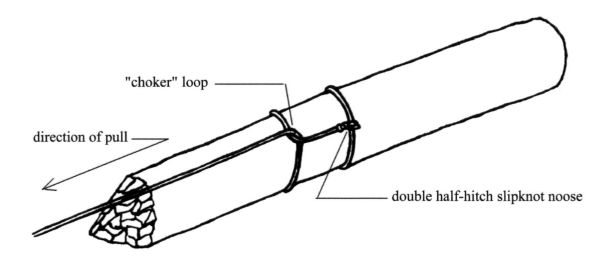

A lumberjack's secret weapon for pulling a smooth-sided log, where a simple slipknot would slide off, a timberhitch, or "choker" knot gets tighter, the harder you pull against it.

Fisherman's Knot

A fisherman's knot might seem to be out of place here, but a knot intended to keep slippery monofilament line from slipping untied is just the thing when you have only braided polyethylene or some other equally slippery rope to work with. The secret of holding power with any knot is that the way line is bent, or tied, around itself offers more friction than can be pulled apart, sometimes enough to break the rope without pulling a knot apart.

With most braided or twisted ropes, achieving that level of friction in a knot is relatively easy, but polyethylene (plastic) ropes, used primarily for inexpensive maritime applications and, especially, monofilament fishing lines, demand a knot that provides exceptional friction when pulled against, or its slippery surfaces will slide over one another until a knot comes untied.

A fisherman's knot is effective because it uses multiple turns to increase its surface area, and therefore the friction that holds it together. The strength provided by these multiple bends is such so that the entire knot can be held together by a single hitch and has holding power that exceeds the tensile strength of whatever line it's fashioned from.

To tie this very strong slipknot, wrap the line through or around whatever needs fastening, then cross the line right over left, or vice-versa, leaving at least a foot of free end to work with, if the knot is made from rope.

Wrap the free end around the working length (according to angling superstition) thirteen times. Then bring the free end down through the initial loop—the loop which is securing whatever is being tied—and then back through the loop that is formed by that action. Pull the free end to snug the knot and slide the wraps, which now resemble a Hangman's Noose, downward to tighten the primary loop.

Making a Knot into a Quick-Release Knot

When I was building a log cabin, using nothing but hand tools, it wasn't enough to tie a good knot that wouldn't come loose while working with very heavy and dangerous loads. The knot had to be easily untie-able, so that the rope could be used again for its next task.

Almost any knot can be made to come untied with just a yank of its free end. In fact, virtually every one of us learns to do just that with our shoes by the time we're in the second grade: The knot used to secure the laces of our shoes and boots is, in fact, a square knot that has been tied to be a quick-release knot. It won't come unfastened when pulled on by a shoe's laces, but it unties with a pull on either of its free ends.

The secret is simple: Just double the line into a loop when making the final bend of almost any knot, be it a bowline or a double half-hitch slipknot. If you can quickly release the last bend in almost any knot, it will come unfastened; and if you double the line, bending it back on itself in the final bend, it will come untied with just a yank on its loose end.

Cables

In retrospect, it was a bad idea. Spring comes late to Lake Superior's Whitefish Point, later than almost anywhere in North America, thanks to the Canadian Clipper, which draws frigid air down directly from the Arctic Circle. Some of the south-facing roads were snow-free, but this road, deep in the shaded woods, was still covered by three feet of very dense, wet "Hardpack" snow.

The deceptively even surface permitted my 4x4, which was something of a local legend for its ability to negotiate rugged terrain, to make it 100 yards, to the first dip in the road. Then my truck's frame came down solidly on snow, but there was nothing under any of its four tires.

Even a cell phone didn't work out here, so road service wasn't an option. Unfortunately this wasn't my first rodeo, so I was more than prepared. I looped a 3/8-inch steel tow cable (rated to work with 4,600 pounds) over the trailer hitch ball, connected it to my cable-type "come-along" hand winch (4,000 pounds) and the other end to a half-inch forestry rope (7,194 pounds). I then put the truck's transmission in neutral and started winching.

The truck rose up and out of the holes its tires had made, then immediately sank down to its frame in a fresh spot. I sighed; this was gonna take a while.

About the third time I reset the winch, the well-used (it should have been replaced) cable attached to the hitch ball snapped with a scary whipping sound. Its broken end sang past my waist, snagging the fabric of my jacket and just removing the zipper pull as it shredded a 6-inch gash into the cloth.

If you're intending to camp, or just cruise, the back country of any large forest, it has always paid to possess the tools that might be needed to remove a storm-felled roadblock. Getting stuck in the woods, with no way out, is a problem that typically doesn't come in just-in-case camping plans.

I was lucky; the damage was nothing that a new zipper and a little sewing couldn't fix. Mishaps like this are more often injurious, even fatal. But they're nearly always avoidable by exercising a little more sense than I showed that day.

You expect a lot out of gear used to haul timber (or trucks), and when you're expecting it to maneuver enough weight to kill you, you owe that equipment some respect. A single broken wire in a cable is reason enough to distrust it. A cable that's frayed should be discarded.

The 3/8-inch tow cable I use should be considered minimum. It would be hard to get too heavy an anchor strap or cable. A *Loop End Cable Assembly*—a steel wire rope with clamped "unserved" loops at either end, ½-inch diameter, or larger, is good.

Straps

So is a heavy nylon tow strap; some of the latter rated for amazing workloads, in excess of 30,000 pounds. No size is too big, if you can work with it—it only matters that it won't break loose.

Fasteners

Which fastener you use is a matter of personal preference, so long as it's rated for at least twice the load you expect to put on it. Snap hooks, with spring-loaded gates that close over a cable, rope, or strap after it has been inserted into the hook, are probably most popular, but pinned clevis hooks are also excellent. I recommend against using carabiner-type snap hooks, as these are usually not rated to the high tensile strengths required for lumberjacking applications, which are probably best thought of in tons, not pounds.

Chains

If you live anywhere near timber country, "logging chain" is a common term for any chain with links larger than a dog chain. Chains are a favorite tie-down among loggers who need to secure many tons of logs together, or on flatbed trailers.

Whenever the wind blew, this tree was touching entry cable, running from the utility pole to the house; this was a very tricky felling job, and not recommended.

If necessary, chains can be tied, whereas steel cables (wire rope), especially larger diameters, resist being knotted. If possible, a chain should not be tied if it can be properly hooked or clipped. If a chain lacks a fastener and must be tied, it should be secured with a good rope-type knot (in other words, a granny knot isn't recommended here, either), then the bitter end secured to another link with a piece of rope threaded through the links, like thread through the eye of a needle, then finally tied. Secured thusly, even a light rope is strong enough to hold a chain tied.

Hand Winch

Building a log cabin homestead far back in the woods, where all construction materials had to be carried in, necessitated a muscle amplifier. In a bygone era, that muscle was supplied by large domesticated animals.

But a mule wasn't feasible for me in the twenty-first century. If anything, I was even more isolated than a trapper or settler of the 1800s, because there were no fellow settlers in the area, no one who was living as I had to. But my need to move and hoist massive weights was the same as it had been for pioneers, so I needed to think out of the box.

Pre-loading a large tree with enough strain to guarantee that it falls exactly where it was intended. The extra 10 minutes spent can save thousands of dollars in damages.

Cable Winches

For me, the solution for dragging 700-pound logs through the swamp and lifting them in place on the cabin's walls was provided by a 2-ton "come-along" manual cable winch. Its 30:1 pull ratio meant that for every pound of effort I placed on the lever, 30 pounds of pull was exerted against the log it was hauling. I could lift a quarter-ton log with the effort that it takes to lift a bag of groceries. Length of pull before the cable needed to be pulled back out, and re-set, was only 6 feet; that meant that pulling a load 18 feet required re-setting the rig three times, but the power was worth the tedium.

That was more than a dozen years ago, and despite considerable hard, frequent use since it was taken out of its box, this winch remains in very good condition. I rely on it often to fell sometimes very large trees precisely where I want them to land. It's an indispensable tool for that job, especially when there's no latitude for error.

Securing a 3-ton come-along winch to an anchor tree, prior to felling.

Pre-loading the target tree, limiting the directions that it can fall, even before the rope is tightened.

Taking a tree down with the rope-and-winch method described here is both safe and precise.

Rope Winches

For lighter loads—a 2-ton cable winch is, in fact, overkill for felling jobs—you might opt for a continuous-feed rope winch. Similar to cable-type come-alongs, rope winches are lighter, with working loads of 750 pounds, but they offer a real advantage of having up to 100 feet of continuous 10:1 pull ratio with ½-inch polyester rope.

The trick to accurately and safely felling a tree isn't so much to control where it falls, but where it cannot *fall.*

Hauling Wood

I think it was a record—for us, at least. My brother-in-law, John, and I cut our winter supply of firewood together every year, and this year we'd fooled too much. White-tailed deer season—a two-week long annual affair that had enough local importance to cause many people to schedule their vacations at that time—was just a three days away, and we only had half the wood that long experience had shown we'd need.

Armed with a three-quarter-ton F-250 Ford pickup, a Remington 40-some CC chainsaw, a permit to cut dead wood in the Jordan Valley section of Michigan's Mackinaw State Forest, and two pairs of new leather gloves, apiece, we started on Friday night. By dusk on Sunday, the two of us had cut and hauled twenty-four face cords of firewood and split and stacked most of it. Both pairs of gloves were holed, but we were good to go when deer season opened at first light Monday morning.

Although not technically a tool, for many people, hauling firewood is as necessary a part of timber-cutting as an axe, and knowing how to do it safely is a must.

* * *

Most folks who cut their own firewood haul it in a pickup truck. Most of these have a factory box, with internal fender wells, knee-high steel walls, and a hinged tailgate that unlatches, and folds down.

Hauling firewood is probably the hardest thing that most people will ever do with their pickup. When I was a kid, and wood, for many of my neighbors, provided the only source of household heat in winter, it was common to have a "wood-hauler," an old, often unlicensed, pickup whose sole job was to haul firewood (some also wore a snowplow in the winter months). Carrying firewood is a tough job—tough on the suspension, and tough on the sheet-metal body panels in especially the box. Expect to acquire dings and dents there, at least, and never just throw wood into the box, or you might well take out the cab's rear window.

A polyethylene bedliner is very much recommended. Heavy-duty, full-bed models do allow heavy sections of firewood to be bounced aboard—but you still need to be careful of the rear window. Full-bed mats are better than no protection, but be warned that exposed metal within the box will surely receive multiple dents—to expect otherwise is like taking swim and expecting to stay dry.

Stacking Wood in a Truck

Wood laid in a box should be oriented parallel with the truck, with butt ends facing front and rear. This is the most stable configuration; it tends not to shift around when driving at highway speeds and it doesn't endanger you with an avalanche of wood when the tailgate is opened.

The first row should be pushed fully forward against the front wall of the box. Lay the first layer across from the left bed wall to the right. Lay the next layer atop it, again, from one wall to the other. Don't worry that layers don't fit the space between bed walls exactly; wood settles and arranges itself as you add weight.

If your truck is rated at 3/4 ton, layers in each row shouldn't exceed four high—about even with the sidewalls of the bed. Wood is heavy, and it's easy to load a truck with more than it can safely bear. If a truck squats down noticeably at its wheel wells, you might already have too much weight on it. If the sidewalls of its rear tires (typical trucks are sold off the lot with passenger car-rated sidewalls) are bulging, remove weight until they don't bulge.

If you have a flatbed, you need to add side racks to contain the load, or—my favorite—cover the entire load with an ample-size cargo net (ProGrip 80"x60", about $40). Tie each side down in at least four places, and the back in at least two places. Observe the same rules about over-loading: re: squatting suspension, bulging tires.

Wedges

One of the most used tools in timber work is a wedge. For felling trees, for splitting firewood "rounds" (the nickname for short, unsplit firewood logs), for splitting off fence rails and rough

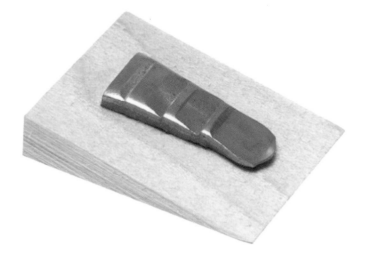

Two types of handle-tightening wedges: the first, a simple, long-sided triangle of hard wood is driven into the center of a tool handle, inside the head, tospread and force its sides against the steel inside of a tool head. This type of edge has been in use for millennia, while the hammer-forged steel model on top of it is less than 2 centuries old.

Steel wedges for tightening loose or worn insert-type tool handles are available at most hardware store for just a couple of dollars.

Easily installed, pound-in tool handle tightening wedges are a must have for any axe owner.

planks from logs that may be of any length, or any diameter, wedges are one of a timber man's most useful tools.

Felling Wedges

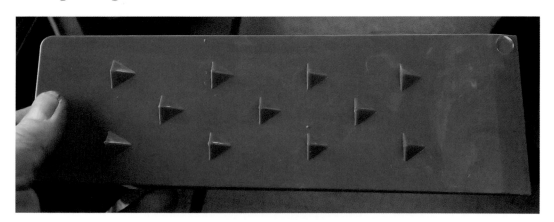

This nylon tree felling wedge has molded-in barbs to keep it from slipping back out of a saw-cut when it's driven in.

Felling wedges, which differ from conventional splitting wedges by being made of nylon instead of steel, are used primarily by arborists and tree-removal companies. They're hammered into chainsaw cuts in standing trees to progressively widen a cut until a tree finally topples over, ideally in the direction and place that a woodcutter wants it to topple.

Felling wedges are also used to forcibly offset a tree that leans in a direction other than the one in which you want it to fall—similar to placing a fold of cardboard under the short leg of a wobbly table. There are limitations to how extreme an angle can be offset with wedges.

Knocking free the cut wedge from a felling notch.

Felling wedges aren't used by loggers and lumberjacks. My local chainsaw expert, Pete, sells them in his store, but he admits that he, like most lumberjacks, professional and amateur (including myself), never use them. In our collective opinions, felling wedges are too slow for felling trees in a forest—these are typically free-felled, because it generally doesn't make much difference where they land. Felling wedges have a tendency to introduce instability, especially on large trees. Finally, they offer no advantage in speed over the methods described here, and are a lot less reliably accurate;

important when the tree you're dropping is growing too near to something that you hold dear. Unless you're a licensed (and insured) tree-removal professional, felling wedges are not recommended.

Beyond that statement, the pros and cons of that argument won't be discussed here. Instead, I'll refer readers to the *Felling a Tree* section a little farther on.

The initial bottom cut to remove a Felling Wedge, which helps to determine in which direction a tree can fall.

The top cut, which angles down to the bottom cut, produces a wedge, which, once removed, forms a notch that helps to determine direction of a fall.

Splitting Wedges

A conventional chisel-type (left) and a "torpedo," or diamond wedge.

Commercially sold splitting wedges are cast or forged of iron or steel. A typical conventional wedge is shaped like a large chisel, beginning at about 3 inches square at its top, and then tapered to a flat point over 6 or 8 inches, weighing 4 to 5 pounds. It's sometimes referred to as a *chisel* wedge.

More recent is the *torpedo* or *diamond* wedge. As its name implies, the shape of this wedge, viewed end-on, is roughly diamond-shaped, or more precisely a + shape. Its four corners offer the very cool advantage of causing firewood rounds (whole logs) to split into quarters. That's the idea but, in truth, doesn't usually happen that way (though it's enough that the halves it makes are already cracked for easy splitting).

The wedge is tapped into the center of a log being split with the flat side of an axe, maul, or sledge, until it stays, then driven down into it with more forceful blows. It is possible to sink a wedge all the way down to its head, embedding the tool, but it's very unlikely that even a tough log can resist splitting before that happens.

Driving a wedge through a log to split it is sometimes the only way to get the job done.

Wooden Wedges

Before there were metal splitting wedges, and even today, in places where steel wedges might be in limited supply, wooden wedges offer effective ways to split long logs—as when making fence rails.

Wooden wedges are a natural product of cutting out felling notches when taking down trees. They obviously aren't strong enough to split wood by themselves, but, if a crack can be started with an axe, maul, or a metal wedge, it can be progressively widened by tapping in wooden wedges until even the largest tree can't resist splitting.

Wooden wedges are best used to split long logs lengthwise, with the trunk lying on the ground. Begin by starting a split at one end, using a metal splitting tool. Then, with that tool still embedded, tap the narrow end of a wedge into the crack right next to the tool on the side facing the log's opposite end.

Ideally, the crack will widen enough to cause the metal tool to fall free. Whether it does or not, tap another wedge into the widening crack farther down the log. Continue to progressively tap-in wooden wedges into the ever-widening crack. Before you reach the other end of the log, it will fall into halves.

Splitting-off wood isn't limited to making fence rails. In Siberia, the short, extra-wide ski-snowshoes that natives use to support themselves on deep snow, or in mucky swamp, are begun in this fashion, as were the rough plank houses of the Chinook tribe, when Lewis and Clark encountered them.

Jack pine can be some of the toughest wood to split, but the Fiskars Super Splitting Axe II, with hammer-forged stainless head, hardened to a Rockwell rating of 55-C (respectable for a skinning knife), with a 36-inch carbon-fiber handle is almost like having a super power.

Splitting a very gnarly piece of jack pine with a torpedo wedge.

In a homestead setting, axes pull multiple duties, like chopping up frozen rabbits for the wolves we kept (under license) for 18 years.

Machetes

Christmas tree farms use a lot of pruning and trimming tools to sculpt their wares into eye-pleasing configurations for some of the most discerning customers in the world. One of the common tools used by workers is a machete.

How sharp a machete needs to be depends on what you're using it for. If you're pruning trees, a machete needs to be as sharp as you can make it, because clean cuts heal faster. But most folks use pruning shears for pruning or a sharp Razor-Tooth® style pruning saw.

In most instances a machete need only have a rough edge, just enough so it doesn't bounce off wood when it strikes. This level of sharpness can be achieved with a rough aluminum-oxide stone or even a mill file, but probably most people find it easiest to just use a coarse pull-through carbide axe and machete sharpener made by the more popular sharpener manufacturers.

For larger wood, with a diameter of more than 2 inches, a saw or a hatchet works best. But if you're clearing brush, taking unwanted limbs from trees, or hacking away vines, the long cutting edge of an oversize knife is often the best tool for the job.

Striking branches is a skill in itself; learn to bring your machete straight down onto a surface being cut. Don't strike at an angle, as this not only diminishes the power delivered, but can result in a dangerous glancing blow—many sometimes serious injuries have resulted from a skipping blade.

My ideal machete is short, with no more than 12 inches of blade, and heavy, with a blade that's at least 1/4- (.250) inch thick and saber-ground to leave as much steel on—and strength in—the blade. Handles should be solid, not hollow, and made from a shock-absorbing rubber or plastic material. Conventional jungle-type machetes with hard, brittle thermoplastic handles, longer 18- to 22-inch blades that are usually .120 inches thick, and made from cutlery-grade 1095 carbon steel simply haven't held up well in a timber-type environment. Repeated hard whacks on wood have broken all of them about 6 inches from the hilt, there's not enough weight to take off heavier branches cleanly, and there's often not enough room to swing a longer blade.

When taking off a branch, aim for the crotch, where it joins the trunk. Strike forcefully, perpendicular to the trunk, and follow through like a golf swing. Trying to halt the momentum of a heavy blade that severs or (more likely) breaks off is hard on your wrist, elbow, and shoulder.

Aside from the potential for joint injury, it's quite tiring, especially if you're trimming branches all day. And fatigue means mistakes. It's frighteningly common for even the fittest person to just throw his or her machete (or hatchet) mid-swing, as its handle slips past tired hand and forearm muscles.

As with a chainsaw, be mindful of what will happen before, during, and after, and never let any part of your body be in line with the blade's arc at any point. And, as with an axe, let the blade and its own inertia do the work. Don't try to muscle it through, because the old warning, "careful you don't take a finger off," should be taken literally when wielding a machete.

Draw Knife

If you mean to use the timber you cut, you might want to peel off its bark first. A lot of traditional cabins (you can see these in old Westerns) were constructed without removing the bark—because building a cabin is back-breaking labor. But bark tends to retain moisture until it finally dries and falls off by itself, making a mess and not doing a thing to preserve the wood beneath it.

A draw knife is a specialized tool made for stripping bark. Not in use much these days, it was once a feature of almost every rural home when I was growing up. The tool consists of a fairly ordinary blade, sometimes curved, to which perpendicular drawing handles are attached at either end. Draw knives were also used as rough planes to shave and flatten or otherwise shape wood, and were a mainstay of backwoods craftsmen. Draw knives are still made today, beginning at about $30.

Brace and Bit

Drilling a hole into wood has always been fundamental to woodworking. From fastening two pieces of wood together by driving a peg through holes drilled into them to just hammering home a peg for hanging coats, drilling holes has been essential since the first homes were built.

A brace-and-bit hand drill was one of the first tools I learned to use in wood shop in school, and it was one of the must-have tools for building a log cabin homestead. A backwoods environment brought home the value of being able to drill a round hole through a piece of wood . . . especially when there's no electrical power.

Bit braces—the tool that holds the drill bit—retail for about $50 to start. Bits average $15 each, a bit more for adjustable-diameter types. This tool has come in handy too many times for me not to have one.

Timber Man

When I was growing up in Northern Michigan, nearly every boy—and a few girls—had learned to operate a chainsaw before reaching legal driving age. The old cast-iron pot-bellied woodstoves that were the only source of heat for many homes had ravenous appetites, and an exceptionally cold or long winter might require forty cords of wood. Running out of firewood mid-winter meant snowshoeing into the woods with a toboggan—or sometimes a car hood—to collect enough wood to get through the next couple of days. Since no one wanted that, everyone worked to help lay in a good supply of firewood every autumn.

The most remarkable chain-sawyer I've known was a man named Dar, who today remains my oldest and dearest friend. Dar actually had the distinction of running his chainsaw for the State of Michigan for a couple of years, and the mimeographed slip of paper he carried proudly in his wallet certified him as a "Licensed Sawyer."

Problem was that he was never very good at it. When it came to safety, Dar was a model lumberjack, always wearing goggles, earplugs, hardhat, gauntlets, and chaps. His technique with a running saw was equally commendable, but when it came to dropping standing timber, my friend had an almost amazing knack for being in the wrong place at the wrong time.

His first mishap occurred while working near a remote power line, when a freak gust of wind blew the tree he was felling onto the wires; he darned near electrocuted himself trying to cut the tree off them. Then, he somehow dropped a large dead elm onto Big John's instep, and it was probably lucky for my friend that the hulking man's broken foot made running very painful.

Dar's wood-cutting saga climaxed on a warm October afternoon in 1983 when he was cutting wood under permit in a section of Mackinaw State Forest near Boyne Falls, Michigan, accompanied by his wife and three-year-old daughter, when his medicine once again turned bad. A tall dead maple he'd notched to fall away from the nearby two-track caught a strong autumn wind and teetered on its stump, torn between the forces of gravity and nature. Dar moved his family to safety and they stood watching as fate decided which way the tree would fall.

But fate can be tricky; just then a Toyota pickup came bouncing toward them in a cloud of dust, headed straight toward the still-tottering maple. Being a conscientious fellow, my friend jumped to the middle of the two-track and began waving his arms wildly as warning for the oncoming driver to stop.

Unfortunately, the fellow driving the pickup was an unsuspecting type. He mistook Dar's gesticulations as a sign that the young family needed assistance, and pulled right up to where my befuddled friend stood under the swaying tree. The man was rolling his window down just as the contest between wind and gravity was decided with a loud crack, and the tree began falling toward the road. The driver's affable grin turned to a look of puzzlement as the fellow he'd thought to help spun suddenly and ran away from him as if pursued by demons.

His questions were answered when a ton of uncut firewood slammed down onto the Toyota's roof like a giant hammer. The little truck's windshield blew outward from the stress of having a tree-size groove stamped into its cab, but the passenger compartment bore up remarkably well. Neither of the two occupants was injured—not physically, anyway—but the vehicle had a definite downward bow to it. You didn't have to be an insurance adjuster to see that this pickup was headed for the scrap heap.

The inevitable lawsuits that followed ended when a judge finally ruled that such an unlikely incident could only have been an act of God. For his part, Dar agreed with the judge, but I've noticed that since that day, he buys all of his firewood already cut.

PROTECTIVE GEAR

You wouldn't play tackle football if you weren't wearing pads and a helmet, and you wouldn't be permitted to bowl if you weren't wearing bowling shoes. It stands to reason that you shouldn't engage in the most dangerous occupation in the world without wearing protective clothing.

Clothing

With a good chainsaw outfit, no tree is too big to fell, no roadblock imprenetrable, but safety gear is a requirement, too.

Often overlooked, proper clothing is a must. I recall seeing a photo posted by a Chicago suburbanite who was demonstrating his penchant for self sufficiency by using a small corded electric chainsaw to cut up a storm-down limb is his backyard. He was wearing short pants and a trendy over-size T-shirt. He was a textbook illustration for unsafe chainsaw use.

When handling unprocessed wood, and the tools used to chop, cut, whittle, or saw that wood, you need only remember that everything is hard, rough, heavy, and sharp.

Not to endorse them, but I personally like *Fire Hose Fabric*™ bib overalls from Duluth Trading Company (www.duluthtrading.com). They're tough enough to offer limited protection from cuts, stabs, and abrasions, and loose enough to permit easy movement, without being so baggy as to snag in moving machinery.

Whatever the brand, trousers should be reinforced at stretch points, double-layered at the knees, preferably six-pocket, with large bellows cargo (thigh) pockets to hold tools and what-nots.

Shirts should always be long-sleeved, with large button-down breast pockets, reinforced at the elbows. Large button-down hip pockets are also good (it's hard to get too many pockets). Probably my favorite work shirt is a GI-style BDU (Battle Dress Uniform) shirt. In truth, whatever it is, it'll probably end up discarded in favor of a T-shirt when the work gets hot and sweaty.

Gloves

Handling wood is hard on the hands. When I had to cut wood for the winter as a young man (not such an imperative these days), I'd wear holes through at least two pairs of new leather gloves each season. It worked out to about one pair of heavy leather gloves for every fifteen to twenty cords of wood.

While old-timers like myself tend to prefer leather, leather gloves are not at all resistant to chainsaw cuts, a fact that I know firsthand. Gloves made from "aramid" fibers, defined most simply as man-made fibers that have a very strong molecular bond and an ability to absorb energy, are better for protecting hands. Nomex® and Kevlar®, introduced by DuPont in 1961 and1972, respectively, are best known of these. Nomex is best known for dissipating heat, but Kevlar is remarkably able to absorb physical forces and is used for protective clothing (DuPont is adamant that

Any tool that cuts wood has no trouble removing fingers, and work gloves, previously made of leather, have been replaced today by the superior protection of aramid-fiber (Kevlar®, Nome®, et al.) handwear.

terms like "Kevlar vest," or "Kevlar gloves" are improper, because DuPont doesn't manufacture vests or gloves).

Aramid fibers similar to Kevlar are Twaron® (Netherlands), Technora® (Japan), and Fenylene® (Russia). Many "cut-resistant" gloves and other protective clothing aren't made from DuPont's Kevlar, but from one of these similar fibers. Manufacturers of gloves intended for working with chainsaws describe them as "Functional Saw Protection" and not "chainsaw-proof," but all of them offer more protection than the heaviest leather.

The danger is that such protection can lull you into an unjustified sense of security. Never regard a chainsaw as anything less than extremely dangerous; the only way to guarantee that you won't be injured is to never allow any part of a bar or chain to touch you.

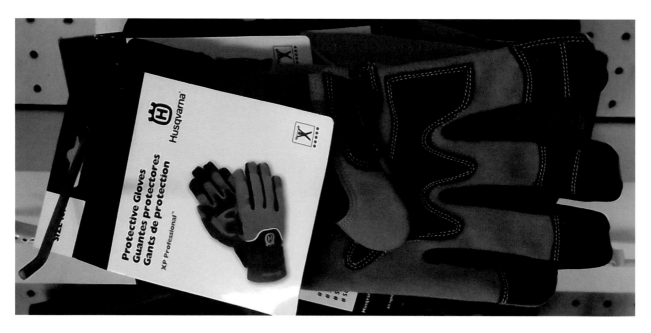

Using a chainsaw requires more than just oil and gas, and a smart chain-sawyer takes advantage of every protection that twenty-first-century technology has to offer.

Boots

Proper footwear can literally be a life-saver, even though boots are probably most often not thought of as safety gear. Mishaps have occurred because layers of dry leaves on a forest floor slide over on another and can be incredibly slippery.

In one instance, a person who was near and dear to me shot his own foot off while squirrel hunting. In another case, a guy who should have known better insisted on hunting in basketball shoes—no one would come near him; he fell on his keister every other step, with a loaded shotgun in his hands. Slipping onto one's butt while grasping a running chainsaw is patently dangerous.

Traditional tall-heeled (made to accommodate tree-climbing spikes), high-ankle Timberman's boots aren't comfortable and, fortunately, they're not necessary for the DIY lumberjack. But lace-up boots with aggressive lug soles that dig in to every terrain definitely are.

So are mid-ankle uppers that extend at least a couple of inches up your calf, past your ankle joint. Sprained ankles are the most common injuries to occur in any outdoor activity, and nowhere are they more common than with lumberjacks who lug heavy weights over rugged terrain. A good pair of working boots is like armor for your feet, and they won't permit ankle joints to over-extend.

Since specialized wilderness-trekking boots first made an appearance in the early '80s (actually, before, but those weren't especially remarkable), their innovative designs have improved the way outdoorsmen of all stripes work and play. Better soles, better support, and Gore-Tex (or some equitable water-resistant micro fiber) will keep your feet dry.

Winter Footwear

In winter, the environment changes and, if you're planning to be outdoors for several hours at a time, your footwear should as well. If there's snow on the ground, you need a "pac-boot"—boot that's built for winter. Insulated hiking boots won't cut it, and it really pays not to find that out the hard way. The main distinction between a summer and a winter boot used to be that one had a removable liner, while the other didn't.

Today, differences have become a little blurred, what with high-efficiency insulations, internal liners, and other innovations. But pac-boots are still rated for winter, snow-country wear, while even insulated hiking boot manufacturers seldom make such claims. Pac-boots are necessarily bulkier, though not so over-sized as the winter boots of even a decade ago. And the generally carry a "comfort rating" (like -20, -40, etcetera), although be aware that comfort ratings are very subjective.

For all-around use, I find a boot rated to be comfortable at -40 to be a good fit for everything. If temperatures actually do drop to even 20 below zero (Fahrenheit), toes will get cold, but not dangerously so and, regardless of theories, it's better to sweat than to freeze.

Good socks are imperative in cold weather. No cotton! That age-old rule holds true. A poor pair of socks can make good boots feel inadequate, while good socks get the most out of even a cheap boot. Cotton absorbs many times its own weight in water, and when it gets wet, it loses all of its insulative qualities.

Straight knit-wool socks are usually too coarse. Pac-boots don't "breathe" well and so are damp from trapped perspiration at the end of the day. In the woods, we'd dry them a bit at night around a campfire but, contrary to some beliefs, you *can* be wet and warm at the same time (e.g., neoprene wet-suits). But damp skin softens, and coarse knit-wool socks can rub the hide right off you if they're worn directly against skin.

I like an acrylic-fiber, non-absorbing liner sock next to my skin. Purpose-made liner socks retail for around $10, but I find the less expensive acrylic "dress" socks sold in department stores for half that price to be equally effective. Ads claim that liner socks "wick" water from your skin, but a more accurate description would be that the heat generated by your foot drives perspiration outward, past the synthetic acrylic fibers, which cannot absorb into the outer sock and the boot insulation, thereby making your foot feel drier and less clammy.

Bug Repellant

When lumberjacking became an industrial occupation, many a strong young man (there were never many old lumberjacks) set out to earn his fortune in the tree-cutting trade. A good many didn't make it, as they contracted a disease known as "swamp madness." In some cases, victims ran wildly through the woods, flailing wildly with axes, and had to be physically restrained. The illness passed a few days after sufferers where taken back to town.

These white spots in the air are mosquitoes, and they pose a genuine hazard to lumberjacks—a mosquito in the eyeball can cause the strongest man to flinch, a dangerous thing.

Swamp madness, as the name implies, was a once arcane affliction caused, for the most part, by exhaustion, which was, itself, caused by unrelenting mosquitoes, blackflies, stable flies, deerflies, horseflies, midges, and ticks. Brutally hard labor, combined with twenty-four-hour torment and no sleep—essentially the same conditions used by Special Forces *Survival, Evasion, Resistance, Escape* (SERE) schools to break the toughest guys in the world—all worked to drive some men insane.

Insect Repellants

Bug protection is a safety feature for lumberjacks—even a hard man may flinch when a blackfly dives into his ear and commences to bite. The most popular commercial sprays contain a chemical

Not often thought of when the conversation turns to lumberjacking, blackflies, mosquitoes and other biting insects can pose genuine safety hazards - a blackfly in your eyeball is sufficient to cause anyone to flinch.

named N,N-diethyl-meta-toluamide (DEET), developed by the US Army in 1946, and approved for general use in 1957. There are approximately 120 DEET insect repellant products in use today, in concentrations ranging from 5 percent to 100 percent, and they're used by about 33 percent of North America to repel biting insects of all species. DEET doesn't kill insects, but makes it hard for biting bugs to detect our odors—we're simply invisible to them. DEET has become the standard for bug-dope, worldwide. It retails in pump, aerosol, lotion, and wipe-on forms for roughly $1 an ounce.

Regardless of a maker's claims, there are moments—like during a mosquito hatch, when literally millions of the insects rise off the water in a cloud—when nothing short of setting oneself afire will deter them from exacting a blood meal. For those times, you need a physical barrier to stop them from reaching you.

Headnets

The best, most enduring protection of all is a physical barrier, and mosquito nets have been providing impenetrable bug protection since lace sheers were invented. The first headnets were homemade affairs, large bags sewn from mosquito netting and just worn over your head.

With the exception that headnets are now available for about $10 at most department stores, the basic design, or their intended purpose, hasn't changed. Wadded up and stuffed in a pocket, there are times when a headnet might be the most appreciated piece of equipment a lumberjack owns.

The clothing you wear can also be important protection against bugs. As already mentioned, baggy clothing with loose folds that might get snagged in machinery is to be avoided, but it needs to roomy enough to allow unrestricted movement, and clothing that slides over your skin as you move doesn't give bugs opportunity to insert a proboscis.

Chaps

Probably the greatest piece of safety equipment since the hardhat, chaps have been a necessary piece of lumberjacking gear since before chainsaws were invented. Before the first aramid fiber (Kevlar®) was created in 1965, *chaparejos* were used to protect legs in rough environments.

Today, protective aramid fibers have undergone many changes—Kevlar alone has been through several generations as gaskets, cables, fiber-optic cables, the infamous "bulletproof" vest and, of course, chainsaw chaps. A half-dozen other companies manufacture similar fibers but, except for minor differences—for example, Elvex Corporation's *Prolar®* material, unlike Kevlar, can be laundered—the fundamentals remain unchanged.

The American Society for Testing and Materials (ASTM) standards, enacted in 2008, require that chainsaw protective chaps be able to stall a cutting chain running at 2,750 inches per minute. Chaps material does this by snagging a chain's cutting teeth, and being dragged into the saw, where it jams the chain and stalls its engine.

Chainsaw chaps do not ensure that you won't be injured by a running chainsaw. A revved-up chainsaw can exceed a chain speed of 4,000 inches per minute, 30 percent greater than speed than a chap is required to stop. Aramid fibers are not chainsaw-proof, and even within its federally mandated parameters (2,750 inches per minute), it will be severely damaged, possibly to the point of needing to be replaced, and you will very likely suffer severe lacerations. The objective of a chainsaw chap is to *lessen* the severity of chainsaw injuries—which typically require about 120 stitches to close—not to prevent them entirely. Never let a chain or bar touch any portion of your anatomy.

The term "widowmaker" has genuine meaning in timber country, where it has been an axiom for centuries; this 24-inch diameter white pine has been snapped off at its base as though it were a pencil.

This cut-off tree went over in a windstorm, roots and all, missing a cabin by thirty feet.

This huge downed white pine, torn up by its roots in gale-force winds,
suffices to demonstrate what might happen to a house or car.

This green tree is nearly 2 feet in diameter, but a lightning strike exploded it as if it were kindling.

FELLING A TREE

It had been a while since I'd seen my crazy friend Larry, but I wasn't too surprised when he slid to a stop in front of me, accompanied by a squeal of tortured rubber. What did surprise me, as he leaned low toward the necessarily open passenger window, was the deep crease in the roof of his once-mint 1968 Ford Bronco.

At the beginning of this book, I described the death of my brother-in-law from a tree. Later, I cited how OSHA has determined that lumberjacking is the most dangerous occupation on Earth. That I'm still alive to write is proof enough that you needn't be a genius to be a successful lumber-jack, but you do need to be careful, and to think before you act.

Most of the danger lies in felling trees. Landing several tons of wood where you want it to go is inherently dangerous. Felling a tree is a controlled crash, not dissimilar to re-entry of a space vehicle or even a bowling ball; you line it up as intelligently as you can, then you let it go, with no further ability to control the end result. You get one shot at it, and it had better be right, because there's not a thing you can do to change where a tree's going once it starts to fall. If you're in the way, you're going to be hurt, at minimum, and if you've misjudged where a tree will hit, the results may be more than catastrophic.

Leave Nothing to Chance

I'd lived alone for two years following a typically awful divorce. I had a vehicle, but my apartment was within walking distance of the forests I loved, in a town that was just urban enough to provide most of amenities of civilization, also within walking distance. Along the stairway that led to my deck grew a beautiful 40-foot mountain ash. At least, I thought that it was beautiful. My landlord objected to the red berry-like fruits that the tree produced each autumn, and especially to the way the smashed fruits stuck to the bottoms of his shoes. Over my—and most other tenants'—objections, he arbitrarily decided that this magnificent old tree must go.

It was no coincidence that I was away for the week, teaching a survival class, when he hired three good ol' boys who'd, literally, never read a book between them in their lives, on how to properly take down a large tree—one that was within touching distance of the stair rail on one side and bordered electrical entry wires on the other.

When I returned from my outing, a week later, the 2x8-inch main rail was still shattered where it had taken a direct hit from the 2-ton trunk, and several of the 2x2-inch supports below it were still in pieces. I ended up repairing the damage, of course, because the hired expert lumberjacks were unlicensed, which could yield no blood or money, and the landlord's insurance didn't cover acts of ignorance.

In the past half-dozen years I've seen as many experts—that is, people who were given money to do the job by a land owner—drop trees directly onto the very cabins, cars, and other objects that they were hired to protect from that tree falling onto them. Where I live, in the timber country of Michigan's Upper Peninsula, the accepted stereotype is that everyone who lives there is a lumberjack. Just like everyone who lives in New Mexico can saddle a horse, or everyone who lives in Chicago must have ties to organized crime. Those images were never factual, especially today, when the video gaming couch potato is generic to every society on Earth.

The technique described here is effective and safe for every felling chore, but always exercise forethought before making the first cut. If you get tired—and lumberjacking is guaranteed to induce that state—stop the activity and go get a beer, or just watch TV, because you're going to make mistakes in that condition, and mistakes can cost more than you want to pay.

Never Try to Free-Fall a Tree

I can drive a tent stake with a falling tree. That's not an idle boast, but rather an honest statement of how accurate I sometimes need to be when felling trees.

So, it's with considerable embarrassment that I relate the following story: As I stated previously, budworm pests are forcing me to fell sometimes large adult trees, and these trees are very often in places where a mistake in precision can result in real property damage.

It wasn't like I hadn't taken down a hundred behemoths before this fateful day, several of them in very tricky spots. It was probably the much bigger jack pine right next to this one that I'd already felled without a hitch that made this dead 10-incher look pretty doable.

Problem was, I hadn't come prepared to fell trees, but only to buck firewood for the pile at home. I had my chainsaw, with gas and oil, and an axe. By my own criterion for must-have tools when felling trees, by my own good advice, presented in these very pages, I was not equipped to fell trees. Especially not my arch-enemy, the always twisted, contrary, and entirely unpredictable jack pine—the only tree I've ever hated with a personal passion.

But, I looked at my wife, Cheanne, and said, "That tree's gotta come down . . . do you trust me?"

She said she did—after all, I *am* the best at this stuff, right?

She pulled the truck ahead to place it farther away than the tree was tall, as you can't be too careful. I revved up my 50cc chainsaw and made a felling notch. And since you can't be too careful (and because jack pines seem to hate me with a vengeance that matches what I feel for them), I made the felling notch point at least a hundred degrees away from the cabin. Just to be sure, I even angled the notch on its lateral plane to point away from the structure. This tree couldn't possibly (you would think) fall toward the cabin.

But fate laughs at the plans of men. I was halfway through the felling cut when the 50-foot tree twisted almost 180 degrees counterclockwise. The trunk pinched my chainsaw and ripped it from my grasp, then actually flung the running saw ten feet through the air. I watched in horror as a tree that could not fall that way, in fact, did. Almost as if in slow motion, the trunk leaned over and collapsed onto the cabin's roof, as precisely as if I'd aimed it there.

I got off lucky. The trunk wrinkled a couple of asphalt shingles and bent the roof's drip edge. Fueled by anger and embarrassment more than strength, I was able to get a shoulder against the trunk and shove it by inches off the hip roof, with no additional damage. Except, that is, for the damage to my oversized ego.

Had I followed my own hard-won wisdom and tied-off and winched the tree the way I describe in these pages, that almost-catastrophe would have been avoided. But I didn't, and I paid for it. The cost wasn't as dear as some of the similar lessons that others have learned, but it very easily could have been. Still, I'll never live it down.

* * *

In the snow country where I grew up, with only a voracious cast-iron, pot-bellied wood stove to keep our house warm, it was a pre-deer season tradition to cut an average of forty cords of firewood every October. In those days, Dutch Elm disease had decimated that once-common species, and we exploited the tragedy by using the hard, hot-burning wood these poor trees provided to heat our home.

In recent years, the Jack Pine Budworm (*Choristoneura pinus*) has been killing off jack pine forests. Victims, ironically, are large adults, 40–60 feet tall, weighing several tons. They are going to fall so, in many cases, they have to be cut to be certain that they won't fall onto something of value or block a roadway.

When I was a boy, timber companies had little use for jack pines. They're the devil's own tree, with soft wood, iron-hard knots, and twisted, gnarled grain that defies efforts to split it as firewood or to mill it into anything resembling lumber. But since Oriented-strand Board (OSB) or "chipboard" was created in 1978, jack pines have had great commercial value to the building industry, and today virtually every structure uses it for roofing and walls. Not all OSB is made from jack pine, but all commercially logged jack pine becomes OSB.

I blame my wife, who's downright dictatorial about where a tree must land; she normally demands more precision than she did in the story that I just related, and for the sake of my own happiness, I deliver. Not long ago, before the near-disaster that I just described, I had a 50-foot jack

The Jack Pine Budworm (Christoneura pinus pinus) is a native insect whose browsing kills large jack pines, making it necessary to fell those that endanger property.   (Photo courtesy USDA.)

pine spin 90 degrees on its stump as it toppled. Had the tree not been belayed, 15 feet up its trunk, to another tree, it would have fallen onto our cabin's roof—and that one was large enough to do some real damage.

Controlling the Direction of Fall

The rope and come-along winch system that I've used since childhood guarantees that a tree cannot fall wrong. If there's anything of value within range of a tree being felled, it's vital that the following procedure be employed. I cannot stress this enough.

Begin by using a ladder to tie a rope rated to no less than 3 tons at least 10 feet up on the target tree's trunk, its opposite end secured to the winch, preferably farther away than the tree is tall. The other end of the winch is secured to a rope or cable tied low around the trunk of an anchor tree.

Draw the pulling rope as taut as you can, double the free end (a doubled rope is easier to untie), and tie it off to the winch hook with at least two hitch knots—one to secure the rope, another to lock the first knot. Then crank the winch snug, just until the winch is held off the ground, in line with the rope—too tight, and the tree will pinch your chainsaw's bar.

Originally designed to be transportable for bow hunters, this lightweight, telescoping ladder has served to fell trees safely and accurately for decades.

Cut the Felling Notch

Almost as much as the rope, this notch determines in which direction a tree will fall, although it is possible for a tree to twist to an unbelievable degree as it falls, regardless of how expertly the notch has been cut. *Never* rely on just a notch to determine in which direction a tree will fall, unless that doesn't make a difference.

Begin with a lower cut, more or less parallel to the ground, which penetrates less than halfway through the tree. Finish with a downward-sloping cut that meets the first. When you withdraw the saw's bar at the apex of either cut, be especially watchful for kickback; I like to rev my saw a little when pulling it free so that the running chain assists in extricating itself, but keep in mind that if you do this not to let your attention lapse for an instant.

With an axe, knock free the wedge created by these two cuts, leaving a notch that, ideally, is perpendicular to the direction you wish the tree to fall. Then, before you make another cut, tighten the winch until the top of the tree being cut begins to sway. With tree pre-loaded in this manner, it cannot fall in the opposite direction and, even it has a spiral grain under its bark, it can twist only a few degrees.

The Felling Cut

The felling cut is made on the opposite side of the trunk. It should be parallel to the notch and begin about 6 inches above it, slanting downward, until it meets the first cut. The slant forms a step behind the trunk in the opposite direction of way the tree is meant to fall and helps to ensure that the trunk cannot slip backward, off its stump, as the tree comes down. (Of course, as we've seen, the best laid plans of mice and men . . . so don't take any chances: Rope it off and use a winch.)

In some instances, when having a small tree under precise control isn't especially important, expert lumberjacks sometimes make a single, downward-sloping cut, at least 45 degrees, beginning at the backside and angling downward toward the intended direction of the fall. The weight of the tree actually lifts off a saw's bar as it cuts downward, falling even as the cut is made, and so, seems to slide off its trunk, leaving a smooth, slanted stump. I do not recommend this technique unless you're, at least, far from anything of value. Be warned that the same twisting phenomenon is likely with a sloping cut, and a tree has often fallen backward, pinching the chainsaw's bar and seriously hazarding the user.

Move!

When felling a tree, be prepared to move smoothly and quickly away; a trunk can sometimes bounce 6 feet in the air.

A tree never falls fast, or without warning, but pay attention, and make sure the area is cleared of tripping hazards, prior to making the first cut.

A tree will start to fall before you can complete this final cut, when the "hinge" of solid wood between wedge and cut becomes too weak to hold the tree upright. So be prepared to move quickly to the side several yards as soon as it starts to lean. A tree always gives plenty of warning and it never falls quickly. But pay attention, as it your life depends on it, because that's exactly the case. Make sure that all tripping hazards are removed from the work area beforehand and that you can with draw swiftly and safely. A little extra time preparing the site is a lot shorter than a hospital stay—or a funeral.

Even though a tree may fall smoothly, you should be at least ten feet away from its butt when it lands. Especially on a hillside, as I've seen the cut end of a 3-ton tree bounce six feet into the air before it came to a rest. Anybody in the path of that bouncing trunk will be swatted like an insect, not metaphorically, but actually.

Widowmakers

A white birch widowmaker, with a hundred pounds of rotting wood suspended 20 feet above ground by a figurative thread of bark, ready to come crashing down at any disturbance.

No matter how carefully you plan, even if you calculate with a pencil, paper, and the ingenuity of da Vinci, a cut tree may be checked in its fall by the branches of a nearby tree. In fact, this is such a common occurrence in a forest that you can pretty much count on it.

Whether a trunk remains on its stump, attached by its hinge, or if it falls off, if the upper branches become entangled enough to hold its upper branches above ground, the tree becomes a proverbial "widowmaker." Never, ever walk under a widowmaker; as the man in the tragic anecdote at the beginning of this article demonstrated: Widowmakers are aptly named.

The correct way to deal with a hung-up widowmaker is to drag it down. The same rope and winch that ensured the tree couldn't fall, except in the desired direction, provides the muscle to drag it free of other limbs. That's why the anchor tree is farther away than the target tree is tall—so that you can't drag it down on top of yourself.

This tree has to be felled, before it comes down with the wind, maybe onto something valuable.

Large dead limbs, and the tops of trees that might snap off and fall, also come under the widow-maker heading. Always look upward when cutting down a tree and be wary of dead limbs and tops that might fall—birch and poplar trees are especially dangerous in this respect.

A few good ol' boy lumberjacks have tried to cut a hanger down by sawing it in half near ground level. *Never* attempt to do this. Whether you're trying to buck a trunk from above or below, it's likely that it'll pinch your saw's bar inextricably. If it does cut through, perhaps a ton or more of wood will slam down uncontrollably (watch your feet!) and, in a best case scenario, the entire tree will become untangled and fall.

This white pine is hung up in the branches of another tree; this situation has killed many an unwary lumberjack: hence, the term widowmaker.

In disregard of the above warning, there will be times when you're forced to cut a trunk that's supported at either end but not the span between. To keep your saw from being pinched as you cut, wedge a block (log) of wood under the span directly beneath where you're intending to halve the trunk. Wedge the block in tightly with a sledge hammer and ore the back side of your axe. Wedged thusly, the block keeps the trunk it supports from collapsing and pinching your saw and enables you to cut clean through the trunk. Remember, one or both ends at either side of a cut are going to roll off the block, so be prepared to pull your saw free quickly and to move your feet.

With the Felling notch removed, this is the cut that causes the (roped-off) tree to fall. First, it will creak against the "hinge" of wood in its center, between the cuts, as it slowly starts to lean, picking up speed as it topples.

Alternately, cut a notch in the top side with side angles of approximately 45 degrees that reaches halfway through the trunk. Knock the resulting wedge free with an axe.

Next, under-buck, that is, cut from below toward the apex of the notch. As the trunk weakens, it will collapse downward under its own weight, opening the cut being made by the chain. Be prepared for kick back, not upward, of course, but straight back, and be prepared to withdraw your saw's bar. I like to support my right forearm—my throttle-trigger side—against my right knee.

Be a Tree Hugger

In many circles, being called a tree hugger is a derogatory term. To a lumberjack, hugging a tree might be a life-saving action. If you hear a cracking of wood or the crashing of branches overhead, looking upward to locate the source might be a dangerous delay. Instead, rush to the nearest, largest tree, and press your body tightly against its trunk. The logic to this maneuver is that live overhead branches help to deflect falling objects.

Note that this is not a theory. I can't relate the numerous times that holding my face and chest tightly against a tree trunk has probably saved me from injury, at the very least. Widowmakers are rarely silent, they make sounds as they're fall to Earth; all you have to keep in mind is that a sound coming from overhead in a tree is never a harbinger of good things to come and may well herald a crushing blow. Short-lived softwoods, like poplar, and especially white birch, are among species of trees to be most wary of.

Felling with an Axe

When I left civilization in 2001 to fulfill my boyhood fantasy of going to the wilderness and building a cabin (the dream of almost every American boy when I was growing up), it was, in retrospect, a bigger undertaking than I'd imagined. In the book I wrote about the experience (*The Log Cabin*, Lyons Press, 2003), I described how, in keeping true to the dream, I used only hand tools—saws, axes, draw knives. This meant that every tree was chopped down, chopped to length, dragged to the building site, and then notched on four places on its ends before being hoisted into place.

Notched and roped, awaiting the final cut.

Chopping down any size tree with an axe will not only endow you with genuine respect for how physically tough and fit were the lumberjacks of old, but with how much skill it takes to put a tree where you want it.

As when using a chainsaw, look before you cut. Study the area. Choose the direction of fall, and make sure (if possible) that there are no branches for a falling tree to become entangled with.

Chopping the Directional Cut

Unlike cutting down a tree with a chainsaw, it's best *not* to tie off a target tree when you make the first cut; a rope will be in the way of your axe swing. It is okay to secure a rope around the tree's trunk at a height well above where you'll be chopping, but don't apply any tension yet, and make sure that the rope is pulled well out off to one side before you begin chopping.

A felling axe needs to be as sharp as it can be. A splitting axe works best when it's a little dull, because it sinks into wood *with* its grain and so tends to bite far more deeply. If an axe is too sharp and it sinks into a block too deeply, it becomes a real chore to disengage it.

But a chopping, or felling axe removes wood chips by cutting horizontally across the grain. It needs to have as keen an edge as possible. By the same token, it needs a blade that is stoutly built, with the weight needed to provide inertia, yet with a cutting edge whose length is relatively short, so that it can break out the chips that must be removed to make one piece of wood into two. For this reason, a broad axe isn't a good choice, except for the dry, seasoned, hand-selected logs that are used in lumberjack competitions. A so-called "race" axe is a poor choice in an actual forest, with green trees, or otherwise wet, knotty wood.

Your first swing "sets" the cut. It should angle downward at roughly a 45-degree angle. The second cut should come in almost horizontally, at a distance below the first, top cut that equals the diameter of the trunk being chopped. This ensures that the progressively narrowing wedge of wood that's being chopped out won't become too closed before you reach the desired depth, about halfway through.

When you make the second, lower cut, disengage the axe with a forceful downward twist of its handle. This action breaks free, or at least loosens the chip between the top and bottom cuts. Don't actually apply your weight to the haft, just twist forcefully; if the chip doesn't break free (sometimes the initial chips are too big to come loose easily), cut the chip in half with another horizontal cut and twist the handle again to free it. The top half of the chip will almost certainly break loose. Then, drive the axe head into the initial horizontal (bottom) cut and twist that half-chip free.

A Note About Competition Axes

Cut-and-Twist is the technique that you'll see universally in lumberjack chopping competitions, as well as in a genuine lumberjacking environment. But take note that axes used competitively by trained athletes bear little similarity to a typical hardware store axe—that is, a working axe that you'll see in the hands of a logger who uses one in the woods to earn a paycheck.

Competition axes are purpose-made for chopping clean, seasoned softwood logs that have straight grain and few or no knots in them. They're heavier, with longer, usually stout, handles and over-sized, broader-beamed heads that weigh 5 pounds or more, and may be custom-made for the contenders who use them (just as a pool hustler may use a custom-made cue, tailored for his preference in bala, thickness, and length).

Unlike typical workaday axes, competition heads are made from high-carbon steel, typically flat-ground, thin enough to be almost fragile and tempered to a Rockwell hardness of more than 55—as hard as some skinning knives. This is mentioned because, while it might seem like this sort of axe is a good choice for everyday work, it is not—the blade will chip if used against knots or hard wood and might be irreparably damaged if a tree should contain embedded wire or nails.

Repeat the next, top cut at a downward angle just inside and parallel to the first. Then repeat the lower, horizontal cut and break that chip loose with a twist of the axe handle. As more and more chips are removed, you'll note that, as with a chainsaw, the overall effect becomes naturally wedge-shaped, becoming narrower the farther your axe penetrates into the trunk and progressively smaller chips are removed. This is why the first "setting" cuts at the top and bottom are spaced so widely; a tree is felled most quickly and efficiently by removing chips that are as large as possible (just watch any lumberjack felling competition and you'll see that assertion proved), and as the wedge-shaped cut gets deeper and narrows, it forces chips to become smaller and smaller.

Securing the Target Tree

When the cut reaches halfway through the trunk, it's time to secure it to an anchor in preparation to felling it. As when felling a tree using a chainsaw, take no chances. Don't allow a tree to do anything that might be harmful to people or property. Use a ladder—a lightweight telescoping type, in the 15-foot range, about $100 retail, is well worth the investment.

For the past two decades I've used a very cool lightweight telescoping ladder, constructed of square steel tubing, originally designed for bow hunting, with a single vertical stay, bordered by three steps on either side; it simply straps to a tree trunk. Three individual sections collapse to a less than 2 feet apiece, telescope to more than 3 feet, and together they allow me to reach heights of more than 15 feet. I cannot find this model anywhere these days, but it has impressed upon me the value of a light, transportable telescoping ladder in when felling trees.

Use the same half-inch diameter, timber-grade rope described previously for felling with a chainsaw (about a 1,700-pound workload). Tie a combination slipknot/choker hitch as high up the trunk as you can reach—with the added muscle of a winch, I've never found it necessary to tie off higher

: A good, strong rope, rated to no less than 2 tons (preferably 3) is necessary to tree-felling.

than 15 feet up, even with trees 2 feet in diameter and 60 feet tall. Tie off the opposite end of the rope to the cable-end snap-hook of a 2- or 3-ton come-along hand winch. With another rope or a double-loop ended cable (rated for at least 3 tons), snap the winch's opposite hook to a solid anchor.

An anchor tree must be stout, at least 8 inches in diameter, green, and able to withstand at least

2 tons of pulling force. It is *not* recommended that you tie off to a vehicle but, if you must, don't exert pull from either front or back, but from a side to avoid putting undue strain on transmission and brakes—the goal isn't to pull down a tree with brute force but rather to anchor a line that guides the tree in the desired direction of fall. If you must anchor to a vehicle or to any heavy object that you value, make certain that the object you anchor to is farther away than the tree is tall.

When the felling wedge has been chopped free of a target tree's trunk, its apex should be perpendicular to the desired direction of the fall—think of it as a rifle sight in that regard. At this point, the rope at the opposite end of your come-along winch is tied to its anchor, usually low on

A strong rope is essential to felling trees accurately.

the trunk of another tree, always farther away than the tree is tall. Finally, the winch is tightened until the top of the tree being felled moves ever so slightly. It's not necessary to exert a strong pull against the tree, and that can, in fact, be dangerous—you don't want to pull the tree down (on top of you), you just want to limit the directions in which it may fall. Once secured, it doesn't matter how twisted or distorted the grain of a tree, it cannot fall anywhere within the 180-degree radius opposite the anchor.

It has been said that the author can drive tent stakes with a 50-foot tree, and the techniques described in this book make that more than an idle boast.

Chopping the Felling Cut

Once the felling wedge is chopped out, the trunk is preloaded under tension from a winch to fall in the desired direction. All that remains is to chop the felling cut on the other side of the trunk, directly opposite, yet slightly above the felling notch. A felling cut severs the strongest fibers in a trunk's outer layers, causing it to lean in the direction of the greatest force, swiveling on the uncut core, called the *hinge* in lumberjacking lingo, until it topples over.

A tree doesn't fall quickly, and it doesn't fall silently. First, it begins to lean, and then the hinge emits a tortured creaking sound, usually accompanied by a *shooshing* sound of rustling foliage from above. Then, if a tree is large, there's that tremendous, ground-shaking thud that has always created a pretty good facsimile of arousal in lumberjacks. Or the tree hangs up in the branches of another tree, which sometimes causes the air around a lumberjack to turn positively blue.

Either way, pay attention every second; if you should find yourself under a hung-up tree or one that begins to topple over on its notch before you intend, there's always plenty of warning and plenty of time to get your body to a safe zone. But you have to pay attention.

Once again, let my late brother-in-law's death serve you: Do not consider any tree-felling task to be a job that you can complete at your leisure; never leave a tree hanging half-cut on its felling notch. And never, ever, walk under a notched or half-fallen widowmaker.

Pulling Down a Widowmaker

The term *widowmaker* predates anyone alive. Like *car wreck* or *house fire*, it was a term that I learned early in boyhood, later the hard way—but, thankfully, usually vicariously. It was, and still is, a common term in timber country where it describes any potentially life-threatening overhead branch: any tree that has been notched to fall or, especially, a partially fallen tree that has become caught in the branches of another tree. The latter is especially dangerous, as it will fall, and it might fall without warning, with just a gust of wind. In my book *The Log Cabin*, I described one such half-fallen tree as a "Sword of Damocles," and that's a fitting description.

A natural widowmaker, a foot-thick jack pine snapped off five feet up its trunk by strong winds; this tree will come down by itself, you just don't know when.

A hanger is going to happen to anyone who fells trees; that's a given, whether you're using an axe or saw, regardless of your level of skill or ingenuity. As often as not, a tree being felled will become caught in the limbs of another tree.

I've seen good ol' boy lumberjacks improvise to bring down widowmakers since I was a kid, and often with disastrous results. The stupidest method I've seen has been an attempt to cut a leaning trunk in half. To do so is incredibly dangerous, not only to the lumberjack, but, if he's using a chainsaw, to the tool, because it will almost certainly be pinched between a partially severed trunk as it begins to collapse under its own weight. A few have gotten away with it; many have not. (The correct, safest method of sawing a tree that's suspended at both ends is the Wedge Method, mentioned previously.)

Another only slightly less dangerous technique is to take down a hung-up tree by also felling the tree with which it's entangled. This, of course, means that two trees are going to fall simultaneously. And they'll fall faster than a single tree, because the second tree is being propelled earthward under the weight of the first tree and it will reach terminal velocity quicker. Taking down two trees in this manner is sometimes necessary, but should otherwise be avoided. If it must be done, exercise all previously described precautions.

The safest and most effective method is to pull a tree backward, away from its stump. This is accomplished by tying a double half-hitch slipknot around the cut (stump) end of the trunk, about 3 feet up its length, then securing it with a timber hitch, which will prevent the slipknot from sliding free. Then, secure the other end of rope to the cable end of a come-along winch and use another rope to anchor the winch's stationary hook to a solid tree, or another suitable anchor. Finally, just use the winch to pull the cut end of the tree trunk backward until it falls free of the branches that are holding it. It seldom takes more than a few feet of pull to disengage a caught tree.

Another very dangerous form of widowmaker is a dead top on a birch, or a softwood tree, such as poplar, aspen, or cottonwood. All of these—and a few other species that haven't been mentioned—tend to break loose at slight tremors to their trunks, sending perhaps a half-ton of dead, brittle wood crashing downward. Birches are particularly notorious for suddenly dropping large sections down onto wood cutters, even though the tough, outer bark shows no sign of the pithy, fragile rot beneath. As described previously, if you so much as hear the sounds of breaking wood overhead, don't look up, and don't waste even a millisecond; wrap both arms around the trunk of the nearest tree and press your cheek against its bark until debris stops falling. This is the safest position, the one that will best keep you from being struck by falling tops.

Limbing

Before a fallen tree can become a log, it needs to have its limbs removed. Most chain-sawyers accomplish this task by revving up their saws and using the tip of the bar to sever limb from trunk at the crotch, where it slants upward from the parent tree. This is a fast and simple process for the professional logger, and by using the tip, they can keep the bar and chain out of dirt that might cause damage to either part. But be acutely aware that a chainsaw is almost guaranteed to kick back when cutting with its tip.

This is not to say that loggers are doing it wrong or that you cannot cut safely with the tip radius of your saw's bar—chainsaw sculptors work extensively with their saws' tips—but be very ready for your saw to kick back. The safest way to trim branches is to place the base of the bar, closest to the motor, top or bottom against the crotch of the limb. But keep in mind that it's also imperative to keep a chain from contacting the earth.

A logger's second choice for taking limbs of any diameter off a trunk is an axe. The objective isn't to cut off the limb, but rather to break it off at its weakest point. This tool requires a bit of skill because, although it can deliver a tremendous blow, its blade is heavy and the striking edge is not very long, so it's easy to miss your mark. Should a misplaced swing bring the cutting edge in contact with your shin or a toe, severe injury will be the result.

In an attempt to avoid hard labor while he was serving a prison sentence, the infamous Clyde Barrow is reputed to have had a fellow inmate chop two toes from his left foot; but there's some doubt as to whether or not that story isn't actually a bid by Barrow to romanticize his own ineptitude with an axe. There is no doubt, however, that an axe can be extremely dangerous.

A heavy machete is easier to aim—and harder to miss with—because of its longer striking edge. A typical jungle machete is too light and too easily broken for lumberjacking work, but there exist today a number of heavy bushwhacking blades that are up to the task. As with an axe, the goal isn't to actually cut a limb free, but to break it loose at is weakest point where it intersects with the trunk. In milled lumber, this limb intersection reveals itself as a knot; striking it forcefully at this crotch usually results in a very clean break.

A little slower and a little less convenient—but a lot safer—is a pruning saw. A generation ago, when the bow saw was state-of-the-art, a small saw wasn't a serious lumberjacking tool. But today, with the new laser-cut designs, a small saw is more than worthwhile. These saws are not only very sharp (be cautious), they cut in both directions, making new ones seem to cut through limbs almost effortlessly. And because they have at least three times the active cutting surfaces of saws made before the latter twentieth century, they remain sharper through extended use. Longer, fixed-blade models are most efficient, but their very keen teeth typically have no sheath; a long-time remedy for me has been to split a section of garden hose, slightly longer than the saw's blade lengthwise, then to slip the cutting edge into the slit, effectively covering the teeth. A section of light rope tied around its middle keeps this makeshift sheath in place.

Topping

I told my wife that the tall red pine was going to have to be felled before we erected the chain-link fence around our kennel's perimeter, but she balks at cutting trees unnecessarily and this healthy 40-footer seemed just too nice a tree to kill.

Lost timbers in Lake Superior, some of them a century old; Great Lakes timber like these fetch a premium price.

Rivers, tributaries, and man-made ditches were used to float logs at least part way to the mill, and a lot of those logs are still there today, stuck in the mud.

Now, a decade later, the 50-foot tree has died from being peed on too many times by our sled dog teams. Its branches, some of them large enough to damage the roofs of doghouses when they hit, had started to fall every time a storm blew by—and on the shore of Lake Superior, there are some pretty impressive gales from time to time. The red pine needed to be felled before it killed one or more of our teams.

Unfortunately, there was no place to fell the tree. There were simply too many immovable—or unfeasibly moved—objects of value surrounding it. The only workable solution was to top it—that is, to climb the tree, secured by a safety harness, then tie off 4-foot sections of its trunk, beginning at its top, before cutting them loose with a chainsaw. Once cut free, sections would fall as far as the rope that secured them permitted before being lowered to Earth.

This procedure is incredibly dangerous. Everything about it is too dangerous for DIY lumberjacks, and the only words of advice that will be offered about tree-topping in this book are to hire a licensed (and insured) professional to do the job.

Bucking

In some places, native and invasive pests make it necessary to cut down trees, before they come down on their own - maybe in a place that you'd rather they didn't.

Once a downed tree is limbed, it needs to be *bucked* to length, in preparation for the lumber mill, barn poles, fence posts, etc. Typical length for commercial logging trailers is in 2-foot increments—for example, logs slated for milling to standard 8-foot-long 2x4 studs (with a little extra length for "trim allowance").

Many a house or cabin, even today, has been sided with rough-cut planks that are simply slabs, about an inch thick, sliced lengthwise from a whole tree trunk to make what is commonly known as *slab-siding*. Especially popular in eras past, this was an inexpensive and low-labor means of siding, especially for cabins, and in some cases a single log can side an entire wall. But first, sections of a tree trunk have to be bucked to length—not always the standard 8 feet.

When bucking a downed tree to usable logs, its top is removed and the base end squared, perpendicular to the trunk—if you're using a saw. If you're chopping the tree in twain with an axe, as I did when building a log cabin, you don't usually bother with making butt ends look nice and square but rather leave them chopped roughly. If you build a cabin this old-fashioned way, you'll understand why—and you'll learn why lumberjacks were so strong.

Using a chainsaw is much preferred, but it's imperative, whether using a power saw or the old-style manual bucksaw, that you keep the cutting surfaces of out the dirt as much as is humanly possible. Dirt and sand are death to a cutting chain and they don't do hand saws much good, either.

When bucking a bib tree whose trunk is suspended off the ground by both ends, taking out a notch gives your saw's bar room to cut, without being pinched between the halves.

In some places, native and invasive pests make it necessary to cut down trees, before they come down on their own——maybe in a place that you'd rather they didn't.

Bucking a tree that's lying on the ground takes care not to run a saw's chain into the dirt.

The most important safety equipment in any lumberjacking chore is the operator.

A big tree needs a big chainsaw.

Lumberjacking can be tough enough to test the strongest of men, but you don't necessarily have to be a tough guy to do it well; more important is to be smart.

CUTTING, STORING, AND BURNING FIREWOOD

In most cases you'll be bucking downed trees and branches to firewood lengths, 16 to 20 inches. It's important that logs be securely elevated when they're cut to desired lengths, for the sake of a saw's cutting surfaces and for the sake of a wood-cutter's back. A traditional way of accomplishing this has been a cantilever-frame log stand known as a *sawbuck*.

If you're at the horse or dog tracks, a sawbuck is slang for a ten-dollar bill, a reference to the support's simplest form, which is just an X-shaped frame made from crossed logs. The shape looks like the Roman numeral for ten, thus a sawbuck became a ten-dollar bill.

A factory-made sawbuck, for sawing-off (bucking) firewood sections from logs. (Photo courtesy 1ekeus.)

In the construction trades, a sawbuck is sometimes synonymous with sawhorse, but to a lumberjack, a sawhorse isn't a sawbuck. The name sawbuck derives from the nineteenth-century Dutch *zaagbok*, and describes a purpose-built stand that holds a log securely while its end hangs free at a convenient height for sawing it off.

Sawbucks have been a feature of wood-burning homes since Biblical times, and there are numerous manufacturers still making them today. Early settlers used to say that every piece of firewood warmed you twice: once during its cutting and once when you burned it. That axiom was in reference to the work it took to saw off firewood lengths with a bow-frame bucksaw, but even in this era of labor-saving power tools, it still requires a bit of muscle to lift logs onto a sawbuck frame.

In practice, most folks who need to cut many cords of firewood at once (as opposed to woodsmen of old who hand-sawed firewood as it was needed) don't bother with a sawbuck. The good ol' boys that cut scores of cords of firewood before winter every year just place their revved-up chainsaws at a comfortable angle across the top of a downed log, then they cut it through *almost* to ground level. When the saw has cut through, with only the merest strand of wood (doing this is an art in itself) connecting it, they give the nearly severed block a hearty push with the sole of their boot, breaking it completely free. They then move down the log's length and almost-cut another section.

This technique is, of course, limited to chainsaw users; bucking wood with a crosscut saw is still best accomplished with logs elevated to about waist-high. As each piece is cut free, the main length is advanced on a sawbuck and the next block sawed off.

Cutting firewood lengths in the woods is where knee-shin-ankle guards that strap over those portions of your anatomy come in handy. Stamped from aluminum or molded from polyethylene plastic, these guards can prevent a lot of bruises and abrasions to the lower legs. They will not stop a throttled-up chainsaw or a full-power axe swing, but they will deflect most of the painful minor injuries that seem to just be a feature of "getting-in the wood."

Splitting Firewood

Cuttin' firewood before winter has been a northwoods tradition for centuries.

Before logs of more than 6 inches in diameter begin to burn well in most wood stoves or fire-places, they need to be split. In some instances, when there's a good layer of red coals in the bottom of a stove's fire box, an entire log of dry, seasoned wood can be added before going to bed at night to "bank" the fire—that is, cause it to smolder slowly and maybe last through the night. This technique isn't so used in these days of precision-controlled burners but, in the era of cast-iron pot-bellies, we employed every trick in the book to make a fire survive for as long as we could.

Logs bigger than half a foot across, usually bucked to 16- to 20-inch lengths (the maximum length for most residential-class wood-burners), are traditionally split into quarters—split in half, then each half is split again. For very large-diameter logs, or when making easily lighted sticks of kindling, quarters might need to be split again.

A century-old line cabin, of the type that used to dot public forests in places like Michigan's Upper Peninsula as a public service to stranded woodsmen.

One of the few operational emergency line cabins still maintained for public use by the federal Forestry Service, this one-room shelter is heated by a small cast-iron woodstove.

Splitting Block

An essential component of wood-splitting is the splitting block. This is simply a length of large-diameter log, preferably bigger around than the logs being split, up-ended on the ground to form a small table of sorts. Logs being split are placed on end atop its upper, preferably flat surface, then struck in their centers with a splitting tool: a maul, splitting wedge, or an axe.

The purpose of a splitting block is to elevate a log being split to a height that's at the apogee of your swing: the point at which a splitting tool imparts maximum force to it. The same log sitting directly against the ground is struck at an angle and its bottom is cushioned by earth, both of which lessen the impact it receives. In addition, striking at a lower point in a swing's arc causes you to bend your back more, which is markedly more tiring.

The exact height of a splitting block varies in accordance with length of the blocks being split and the height of a person swinging. Proper height is one that brings the head of your splitting tool into contact with wood when the tool handle is perpendicular to your body, at about waist level.

Reading the Grain

One trick of splitting is to hit a block in the spot at which its structure is weakest. This is indeed a trick, as you cannot see inside the wood, so you have no way of positively knowing how or if it twists inside, causing it to split in a spiral fashion—which is to say that it'll be a proverbial bugger to split.

Some trees grow straight and true, with even grain that causes them to split easily. These are selected for movie stars to split with a single swing. They're identifiable as having a nice round-ness, with evenly spaced growth rings, and few or no knots—knots are left from limbs growing out from the trunk's length. Vertical grains split to encircle places where limbs were, and these places—knots—might make for beautiful furniture, but it's very hard to split surrounding wood.

Heavy snows cause young trees to grow twisted and bent.

Some woods, like jackpine and red oak, are naturally knotty with coarse, spiral grains that make them a nightmare to split. But all trees can become twisted when subjected to external stresses. Trees that are bent over from the weight of snow every winter can be gnarled to almost unbeliev-able extremes. The same goes with trees that grab with their roots for any solid footing in soft, wet swamp (or those that must grow in the face of constant prevailing winds).

When splitting firewood, there's often a natural crack to indicate where the grain is weakest.

Can you tell where this log should be struck with a maul or wedge to split it most easily?

Splitting wood is an art in itself.

Although it isn't obvious, usually, some trees, like this jack pine, have twisted, hard-to-split grain, as a result of being bent under the weight of snow for many consecutive winters.

To burn most efficiently—or to even fit into most woodstoves—large logs need to be split into quarters, or even eighths.

Due to the direction of the earth's rotation, prevailing winds blow west to east. In the north woods, winds tend to blow more from the north in winter, in accordance with the planet tipping on its axis, and cold nor'westers are infamous in timber country throughout the northern US and Canada. This phenomenon has always been important to lumberjacks, as nor'westers tend to kill or stunt branches that face into them on the tallest trees. This creates a rudimentary compass because the thickest, fullest branches on tall pines always face southeast.

Very often, especially on twisted, knotty lengths that you know aren't going to split easily or straight, a block will telegraph its weakest spot to you with a crooked, usually darker line that extends from its center to its outermost ring. Striking that point with an axe or maul is likely to create a crack that roughly follows that line.

In a perfect world, where all blocks of wood have perfectly straight grain, every block splits with a single blow. In a real world that seldom happens. So avoid knots, look for natural weak spots in the grain (like the aforementioned lines), and follow the natural splits caused by a strike with successive strikes in exactly the same spot; the latter precision is a skill that's practically impossible to achieve, but it's a goal that every axe swinger should aspire to. An accurate swing is also a pretty handy thing to own for the next section of this book.

Splitting Wedges

A splitting wedge is, as its name implies, a wedge most often made from steel these days or sometimes malleable iron that is driven into the top of, then through, a flat-ended log of firewood to split it into two evenly spaced (usually ideally) halves. A splitting wedge may be employed again to split halves into quarters although, once a block is halved, it's generally weakened enough to split easily using just a maul or the latest generation Fiskars-type splitting axe (I swear, I have no affiliation with the company, but using that axe to split with is almost like having a super power).

In its simplest form, a wedge is not much different than an extra-wide railroad spike, almost a chisel—both of which have, in fact, served as impromptu wood splitting wedges, at logging, mining, and other quickly set-up camps in the past. A proper splitting wedge is larger than an average man could manage as a chisel, which it otherwise resembles. The alloys, homogeneity (absence of overly soft or hard spots) of, usually, hammer-forged steels of today makes splitting wedges of yesteryear seem poorly made by comparison.

A large splitting wedge is, by itself, equal to all but the biggest-diameter and most gnarly grained firewood-length (16 to 20-inch) logs. There are various sized splitting wedges; mostly spike-shaped, but sometimes diamond-shaped, 3 to 5 pounds heavy, 7 to almost 10 inches long, usually quite blunt (I like to sharpen a steeper cutting edge onto some of mine) and about 2 inches wide on all four sides.

Smaller splitting wedges, are, of course, made for smaller splitting chores, but smaller usually sharper wedges are also easier to start into an un-split block, and the crack they begin accommodates the larger wedge, which finishes the job with minimal effort. It's of no matter if you have two wedges embedded at the same time, and the smaller wedge will likely fall free anyway, as the block splits. Few firewood-size logs can resist the separating forces of 9 inches of 2-inch wide and thick steel being driven through their fibers.

Torpedo, or diamond wedges, have four tapered sides and are arguably superior to the traditional chisel wedge. They're generally shorter—a 4-pound model is just over 7 inches long, but wider than their conventional counterparts. Their greatest advantage, aside from usually being a lot easier to tap into place, is that they divide a block in four directions instead of two. In an ideal world, they'll split a block into quarters. In real life that rarely occurs, but it certainly doesn't make cloven halves more difficult to split.

Few axe makers warrant the backs of their single-edge axes as sledges, and if you're going to spend a day splitting blocks of pre-cut firewood-length wood, by all means, you should use a proper sledge hammer and a proper wedge. A hammer is heavier, and therefore more effective, and there are a few lighter axes that are just too lightly built to act as either hammer or wedge. However, I have an axe that I've been using hard for both those two purposes for almost two decades and, despite being on its third handle, it has a lot of years left in it.

Splitting with an Axe

Splitting a winter's supply of firewood with an axe is never recommended, although doing just that, often accompanied by a sledge hammer and splitting wedge, was the norm for most of past centuries. An axe's blade is too narrow to separate tough wood far enough to force it to split cleanly and it lacks the weight you really need to drive its edge through a block.

These two drawbacks make for a tool that tends to become firmly embedded in wood with every stroke. Loosening a stuck blade necessitates tipping a block onto its side, whacking the axe handle with the heel of your hand to loosen it, and then levering its handle up and down until the bit comes free. Some types of wood—maple, poplar, cottonwood, etc.—in low northern woods (where they aren't twisted and distorted each year by snowfall deep enough to encase them) split easily and cleanly; you see a lot of these being split by movie stars but, in real life, don't expect blocks of wood to split so readily.

An axe's narrowness also causes its blade to sink into wood, sometimes down to its handle, without parting the block into which it's sunk. This almost unavoidable problem—the cause of most broken axe handles—causes an axe haft to strike against the block, splintering the handle just below the tool's head. Enough strikes like this and a handle breaks in half, possibly causing its sharpened steel head to go flying dangerously through the air.

But a lot of firewood has been split with an axe because it had to be. Splitting mauls, which had been hand-made by farmers and homesteaders for decades, didn't make a commercial appearance until the 1970s and, even though mauls are common today, they're single-purpose tools. They don't work well for felling trees or chopping wood.

But settling a frontier doesn't permit the luxury of hauling along any more specialized tools than are necessary. Frontiersmen of old were limited in the weight they transported into backcountry. An axe had to pull as many multiple duties as it could.

A double-bit was the favorite of lumberjacks for whom the same tool had to chop down standing trees, then split or otherwise transform those trees into usable forms. It was common to sharpen one side to a very keen edge that really took a bite when chopping (or sometimes even worked as a skinning knife, for large game), but to leave its opposite edge fairly dull so that it wouldn't embed itself so deeply when splitting firewood or fence rails.

A major drawback of the double-bit axe was that it couldn't be used as a wedge, pounded from its opposite side, down through wood the way a flat-backed single-bit could. Most axe-makers discourage this practice today, but an awful lot of wood has been split using two single-but axes. If one gets embedded tightly into wood, the other's flat back is used as a sledge to drive it down through. Be aware that there are some light-duty, usually (ironically) survival-type axes whose design isn't beefy enough to be struck as a splitting wedge. Some clearly can but, to be safe, try not to use an axe as a splitting wedge; get a dedicated splitting wedge for around $10. They may peen over a little at their tops over the years, but modern hammer-forged steel splitting wedges are essentially indestructible

Splitting Rails

In the early days of civilization, any growing nation needed farmers who reared and raised livestock demanded fences. Fences of the day needed split rails. Lots of them. So there was plenty of work to be had for a good rail-splitter.

The conventional method for splitting long (8 feet or more) logs into fence rails, even crude planks, is to lay a log on the ground and then chock it in several places along its sides to keep them from rolling. Starting at one end, drive your axe lengthwise into its center, causing a short split to develop.

Drive a wedge into that split. A steel splitting wedge is most ideal but, in early days, the wedge was a simple triangle, even just a chip of wood, anything that could be pounded in place to maintain or widen the beginning of a crack. No need to drive the wedge more than half an inch or so—just enough to keep a split growing ever wider with each wedge.

Next, drive another wedge into the crack that's developing toward the log's middle. This widens the split and causes it to lengthen, reaching toward the log's opposite end.

Another wedge driven into the crack, at its farthest point, forces the log to split farther.

And so on. With each wedge, the wedges driven in prior will become looser, as the crack they sit in becomes wider. Lightly tap them down until they become snug again. But keep adding wedges until you reach the opposite end, as more wedges help to control a split and keep it straight.

In no time, even a very large log will (usually) split neatly in half. An adze, or a large draw knife (or even a small axe), allows you to smooth both sides and turn split trunks into floorboards or tables—surfaces that need to be at least somewhat flat. With only a few hand and measuring tools (and a bit of ingenuity), you can render trees into usable lumber and from there into furniture, houses, water-wheels, etc. Remember that civilization was born from wood.

In Siberia, even in modern times, 3-foot lengths of such rough, split-off planks about a foot wide, are smoothed with a hatchet to a thickness of about 1 inch. Then (depending on the expertise of a craftsman), the ends are made pliable by heating them in boiling water, giving them an upward bend in a jig—which is also crafted of wood. When these ski-shoes are shaped and dried, they're drilled out to accept a rudimentary rope-binding and are used as a hybrid ski/snowshoe for traveling atop deep snow or through mucky marsh.

Smaller diameter logs are split into quarters and used as fence rails. One of Abraham Lincoln's first jobs as a young man was splitting rails, and there may be a causal relationship between that hard way to earn a dollar and Honest Abe's legendary strength of character. In a world where lumberjacks and the craftsman to whom they provided timber transformed vast, old forests into houses, wagons, boats, even windmills, anything seemed possible.

Making an Axe Handle

I recently heard a lady refer to one elderly fellow's axe as "well cared for," with a wood handle that seemed to be "polished from sweat and hard work." My unintentionally cynical reply was that it likely hadn't seen much work at all.

I mean, come on, this is a tool—like a pick axe or sledge hammer—that has been intentionally designed to strike forcefully against other solid objects. No tool is used harder than an axe so, unless it isn't being used as intended, it's pretty normal to break an axe handle once in a while.

I go through about two wooden axe handles a year, victims of hammering and twisting that goes with chopping wood and, around here, the hard impacts associated with chopping blocks of granite-hard frozen venison scraps, first for the wolves we raised under license for eighteen years and, after they'd passed from old age, our sled dogs. The level of health and the longevity enjoyed by all of our wolves and dogs for decades is proof enough to us that a red-meat diet is best for canids.

You can buy replacement axe handles at almost any hardware store for about $15, but I live in the woods, so making the trip to town is not only inconvenient but kind of redundant. People living in the backwoods, even today, have always made their own axe and other tool handles.

Broken axe handles are part of working life, and where an axe is most valuable, there often isn't a hardware store handy. The first step to carving out a suitable replacement is to cut a sapling large enough to fit the bill.

Whittling down the wide end of a sapling to match the recess in an axe's head (using a short, heavy SP-8 machete from Ontario Knives).

Fitting the freshly hewn axe-handle into its head.

The rough-finished handle seated in its head, ready for work (or for smoothing, and prettying).

Generally, I just saw off a hardwood sapling (I've made some pretty good handles from oak and—my favorite—maple). Preferred saplings are without knots and with a generally straight grain. A candidate for an axe head should be 6 inches in diameter; you need it to be as large around as the axe head's teardrop (single-bit) or almond-shaped (double-bit) head slot is long.

Whittle the end to match the shape of the handle slot in the head. For the coarse work, where a lot of wood needs to be removed, I prefer to use a large, heavy machete—even a sharp hatchet. Chop the blade into the wood, and then wiggle it downward under pressure until the desired amount of wood is removed. It pays to take smaller bites, as these are easier to remove and because it's easy to remove too much wood—and it's harder to put that wood back. Removing too much wood isn't necessarily catastrophic, though, because strategically hammered-in wooden shims aided by a little expanding glue, like Gorilla Glue™, about $6 at most hardware stores, fills in gaps.

I prefer to leave the end that inserts into the axe head just a little larger at the bottom of the head than the hole into which it fits—just a little: with a slight flare at the bottom of the inserted portion to prevent the head from going down too far onto the haft.

Seating a head in place onto the new handle you've crafted can be accomplished by lightly pounding over its whittled insert, then pounding the opposite, bottom end of the shaft against a solid surface (your splitting block works great) with deliberate, forceful blows. The forces of inertia and gravity work to drive the head downward onto its handle.

If the handle is too tight, this is one of the few jobs in which a guy can be a guy and get physical, placing a foot-long section of 2x4 onto the head with the bottom of the handle braced against the top of the splitting block and beating hell out of it with ball-peen hammer or a small sledge hammer (even a wooden club) until the parts fit.

Conversely, if you remove a bit too much wood and the handle fit is a little too loose, that problem is easily remedied by driving one or two wedge-tightening, serrated metal shims into the topmost surface to expand wooden sides tightly against metal from the inside. These are sold, usually two to a pack for less than $4 each.

Once the head is mounted solidly and securely into the new axe, all that remains is to whittle the rest of the haft until it matches your personal wants and needs. Whittle here, whittle there, and you've created the ideal axe for you, perfectly suited to your body size, your strength, and to your personal grip. For me, a perfect hand-crafted axe handle is a little fatter, more hand-filling, and at least 36 inches long to give it a lot of inertia and striking power.

Splitting with a Maul

Better than even a Timbercruiser-size (5–6 pounds) axe when it comes to splitting wood is a splitting maul. Not available unless you made your own but, when I was a boy splitting wood every night for our voracious pot-belly this tool is almost invaluable to those who need one. With a heavier, wider tapered head than an axe, it drives itself deep into a block without getting stuck and forces even gnarly wood grains to separate.

Mauls typically come with heads weighing 6, 8, and 12 pounds. Still at the log cabin I built for my book, *The Log Cabin*, is a homemade maul with a massive 15-pound head. That maul splits whatever you hit with it, but it isn't for the faint of heart.

Splitting firewood with a maul is easier than it is with an axe, even though the tool is two to four times as heavy and for the same reason that a heavier axe head can, in fact, chop wood quicker

This 12-pond maul is what is known colloquially as a "monster maul," and there aren't many pieces of wood that it can't split. (photo courtesy of Northern Tool company).

and with less expenditure of energy than a lighter-headed axe. The secret to using a maul—as with an axe—is to let the tool do the work. Do not try to power it through a log, because attempting to do so is guaranteed to reduce a world-class athlete to a sweaty mass of quivering muscle after only a few swings.

A splitting block made from a large upended log, about 2 feet long, is recommended. Just as it is when splitting with an axe, because elevating the wood you're splitting puts it at the most efficacious point in your swing and is much easier on your back.

Unlike an axe, the starting point for a swing should begin with the maul's head resting on the ground, not supported over one shoulder. This might not seem significant in terms of saving energy, but it is. Don't lift the maul any more or higher than you have to. Hand placement on the haft is the same as it is with an axe, with your strongest hand nearest the head and your weaker hand near the handle's end.

With the maul's head only slightly off the ground, use your strong arm (its hand gripping the handle) just below the head to rotate the maul backward in an arc. As it rotates through its motion, your strong hand slides toward the middle of the shaft. When the maul's head is vertical, that hand should have slid at least half the handle's length toward its end to rest against the top of the other hand.

Again, let the tool do the work, not your muscles. The heavy, sometimes massive, head on the maul only seems like it takes a lot of strength to operate, but many a teenager—even young girls—can make a splitting maul work for them. Use only as much power as it takes to guide it as you swing it around in a circle to make contact with wood. As described above, terminate your swing while hanging onto just the last few inches of the handle's end, allowing the head to make contact under its own inertial power.

Hydraulic Splitters

Even if you're very skilled at reading wood grain and determining where a log's weakest point lies, and even if you have the world's most complete mastery of a splitting maul, it's still very hard work to split logs using a striking tool and wedges.

If the job at hand is a large one—like getting in a winter's supply of firewood—the hydraulic muscle of a wood-splitter is an easier answer. No one really knows when these were originally invented or by whom, because the first units were produced in the barns of good old boy farmers who possessed a stick (carbon-arc) welder and the boatload of ingenuity born of the necessity-driven inventiveness that has always been a trademark of rural dwellers.

Today there exists a bevy of wood-splitting machines from numerous manufacturers. The first—and still most common—wood-splitters employ a hydraulic ram that amplifies a smaller amount of force to drive a large steel wedge through (usually) the flat, cut end of a whole log. Typically, a single-blade wedge halves the log, and then the halves are split again into quarters. More recently, four-cornered diamond wedges have appeared that split a log into four equal sections.

A very old hydraulic-ram wood-splitter that was built in someone's barn.

Now available from numerous manufacturers, in a variety of configurations, even powered by electricity, the first hydraulic wood-splitters were fabricated by good ol' boys with stick welders.

The hydraulic pistons that force splitting wedges through wood are traditionally powered by a two-stroke piston engine. More recently, splitting machines are driven by a plug-in electric motor that's powered by standard 110 volts AC from an ordinary household outlet.

Another recent wood-splitting device is a giant conical screw that literally drives itself into a chunk of wood, usually from the side instead of one end, using a rotary force. As the screw bites into wood, it becomes increasingly larger in diameter, forcing even the toughest grains to split apart.

The concept is great but, in actual practice, current screw-splitter designs are a bit unwieldy. The most inconvenient of these derives from an automobile's axle, using torque provided by the car's wheel hub to drive the screw. This involves removing one wheel, jacking the entire drive axle up, fastening a screw-splitter to the wheel hubs (questionable, in itself), and running the car in-gear.

Other types use PTO (Power Take-Off) drives from tractors and even high-torque electric motors. I think the screw-type wood-splitter is great in theory, but I feel as though the practice needs refinement.

Storing Firewood

We cut a lot of green trees around here thanks to bark beetles, fungi, budworms, and other parasites. Healthy trees are usually safe from us, but many trees are in the slow process of dying when we cut them. Still full of sap, some of which is extremely resinous when burned, green wood needs to be "seasoned" to serve as firewood or else it'll coat stove pipe and chimney walls with dangerously flammable creosote—a tar-like substance that can, in some instances, explode.

For that reason, green wood needs to be stored properly until the moisture in it dries, and it weathers sufficiently to burn as hot and as efficiently as it can, but without losing those properties of combustion to decay or dry-rot.

What Is a Cord?

The Merriam-Webster dictionary describes a cord of wood as: "a unit of wood cut for fuel equal to a stack 4 x 4 x 8 feet or 128 cubic feet." In fact, firewood isn't generally cut to 4-foot lengths, because a log 4-foot long won't fit in a wood stove.

For that reason, firewood is sold by the "face cord," or "Rick." It's called a Face Cor because, if you look at the stack two-dimensionally (or on its face), it measures 4 feet high by 8 feet long. But individual pieces of firewood are cut to more stove-friendly 16- to 20-inch lengths. The actual volume of a face cord of firewood is approximately one-third of a full cord, or about 43 cubic feet. This will, of course, vary depending on how long individual pieces of firewood have been cut—a bit more if lengths are 18 inches; even more if lengths are 20 inches and so on.

Essentially, a cord is nothing more than a useful unit of measure—you've got to have something real to use as a reference. Depending on how large the space you're heating, how well insulated that space might be, or how cold or long winters are (as well as other factors), the number of cords you'll need to heat your home might vary considerably. When I was a kid and wood-burning heaters usually took the form of giant furnaces with octopus ductwork or, more usually, cast-iron pot-bellied stoves that often glowed red on especially cold winter nights, they might consume more than forty face cords. Adding a bowling ball—size lump of slow-burning coal before bedtime helped to decrease consumption, but the advent of "airtight" stoves in the 1970s halved the volume of wood needed, and today's high-efficiency wood burners have cut that even more.

As a rough rule of thumb, presuming a home is being heated with a high-efficiency burner, I figure on needing ten cords of wood for each bedroom in a house. Two bedrooms, twenty cords of wood. Again, that'll vary, according to a number of factors, but, like a face cord itself, that's an approximation.

Types of Firewood

Fundamentally, there are two types of wood: hardwood and softwood. All coniferous (evergreen and needle-bearing) trees—pines, cedars, tamaracks—are softwood. All hardwoods are deciduous (leaf-bearing) trees; examples are maple, oak, beech, and elm. But some deciduous trees—like cottonwood, aspen, and poplar—are softwoods. Still others, especially white, silver, and yellow birches, fall between those two groups.

Generally, hardwoods are coveted for making sturdy, long-lasting furniture, although the beautiful grains and availability of fast-growing pines has made them a choice for cabinetry and other furnishings, too. The density of hardwoods makes them the first choice for home heating, as they tend to burn slowly and at a high temperature. Softwoods are known colloquially among rural folk as "gopher wood," because you fill the stove, then "go fer more."

When forests were more plentiful and varied than they are now, softwoods were shunned, even disparaged. Today, hardwoods are rarer, and we can ill afford to be so picky on any level. Modern woodstove designs get maximum BTUs from every type of wood with cleaner, more thorough combustion than ever before. White-bark birch logs—the traditional Yule Log—fetch premium prices for fireplace displays at high-end hotels and the like, but the average consumer who just wants help with heating bills can safely burn any species of tree.

Curing Firewood

The secret to getting the most from your firewood has always been to *season* it well. That term derives from the practice of storing freshly cut wood for a season—translated in this case as a summer solstice—before burning it. Green wood of any type is hard to ignite, as it burns poorly and incompletely and the high moisture content of its sap tends to form a tar (creosote) that coats chimney liners with a dangerously combustible layer.

Seasoning wood isn't drying it, exactly; seasoned wood that gets wet doesn't have the same characteristics as green wood. Seasoning is a curing process that coagulates and elementally changes the structure of wood, much like ripening a fruit before it becomes edible.

Stacking Firewood

How you stack your firewood while it's curing is important. An awful lot of folks simply have the cut-and-split wood they buy just off-loaded into a small mountain in some unused corner, and then take it as they need it as they need it throughout the winter.

This is a bad idea, even if the wood has already been seasoned (it generally is). Mounded lengths of wood don't breathe properly—that is, air doesn't circulate evenly throughout the pile—and the interior of a pile retains moisture. If the wood isn't used within a few months, pieces at the center of the pile, especially those in contact with the earth, will begin to rot.

Properly stacked and stored firewood can remain useful for years. Of primary importance is keeping the bottom layer off the ground. This can usually be accomplished by laying a pair of, say 2x4s parallel on the ground liked railroad tracks, then laying split firewood across them so that both ends of the wood are supported and held off the ground. Supports don't have to be 2x4s, they can be any boards, patio blocks, even lengths of firewood laid end-to-end in a railroad track configuration. The objective is only to ensure that stacked wood is held even slightly off the earth so that it does not absorb moisture that can cause it to decay, thereby removing support and causing the entire stack to collapse.

Protection from rain and snow isn't usually a necessity, but most folks like to cover their woodpile to keep top layers from getting wet or to keep them from being buried under deep snow. A polyethylene tarp, even a plastic painter's drop cloth, available inexpensively from most hardware stores, is sufficient, although in heavy snow climates it should be thick and tough enough not to rip when you shake ice and accumulated snow from it.

Be aware that covering your stacked wood with a waterproof cover can be more harmful than good, especially in rain forest—type climates. A watertight sheet means that moisture from above is kept out, but it also means the moisture below the tarp is trapped and unable to evaporate. This can cause wood to molder, rot, and be wet enough to resist combusting cleanly in a stove. When using a watertight blanket of any type to cover your woodpile, it's best to cover only the top layer, allowing air to circulate freely among lower layers.

A minor digression here, if I might: In snow country, where winters are frozen, not wet, wood doesn't rot. Frozen water is ice, and isn't wet but rather dry. For it to be wet, water first has to reach above 32 degrees Fahrenheit. This is important to know, because a an improperly piled mountain of wood in Seattle, Washington, or Roanoke, Virginia, may be destroyed by rot in its center before spring, while a pile in Marquette, Michigan, remains unchanged.

Then there's the classic woodshed. In most cases, this was, and is, a misnomer. A woodshed typically has only a roof supported by a frame of lumber, usually a single-plane sloping roof, lean-to fashion, but taller. Precipitation is shed and lack of walls permits free circulation of air. In generations past, when some say that kids grew up to become better-mannered men and women than they are today, being "taken to the woodshed" was more than a figure of speech.

An ability to fell trees, buck, buck logs, and notch poles enables a woodcutter to provide his own raw materials for building barns and other structures.

Dangers in a Woodpile

Be aware that there may be dangers in a wood pile. Stacked wood is a favorite nesting place for most spiders—most importantly black widows and brown recluse species, as well as scorpions, where these arachnids are native.

More likely, in some places, bees and wasps are known to be found in similar piles. Very aggressive yellow jackets (these guys make "killer bees" look life houseflies) are infamous for making stacks of wood their homes, but don't be surprised to find that any species of honeybee, carpenter bees, wasps, or hornets have taken up residence there. Nests can literally appear overnight, so exercise caution whenever collecting an armload of fuel. Wear sturdy, preferably leather work gloves when handling wood; keep a pair handy and close to your wood pile, where you'll have no excuse not to slip them on whenever you grab a few lengths to feed your burner.

Also make sure to remember the old axiom about "a snake in the woodpile." Several species of snakes, including most rattlesnakes, are partial to the shelter and shade provided by stacks of wood. Snakes aren't overly aggressive usually, but not noticing a copperhead who thinks that your tearing down what it thought was a safe haven can prompt it to strike from fear.

Burning Firewood

Building a fire isn't rocket science, but the ability to coax a stable building material into becoming a self-sustaining blaze on demand is, indeed, a science.

There isn't a "right" or "wrong" way to make fire, but rather only techniques that do or do not work, depending on a multitude of variables. It's an old saw among expert woodsmen that they don't have a way to make fire, they have dozens of ways. They have to, because a few of the most experienced (and getting such experiences is often not an enviable thing) have learned that building a warm fire is sometimes the difference between living and that other thing. And they've learned the need for fire increases in proportion with the difficulty they'll have getting one started.

Tinder

Every fire starts with tinder, because no fire begins as a blaze. When building a fire, "build" is the operative word; you start small, and you progressively add larger pieces of fuel as fire grows hotter.

At its simplest, tinder is a very dry ("Dry as a tinder box," as the old saying goes), very fine, very easily lighted material that will catch fire with the touch of a match flame or even a hot spark. In the old days of flint-and-steel, every frontier child knew how to take a "bird's nest" of fibrous bark, "char cloth" made from singed cotton flannel material, powder made from dry-rotted wood, or even super-fine dry wood shavings whittled from a solid piece of wood and start any of them to at least smoldering with just a spark. (In fact, it was kind of important, because, in some households, allowing the hearth to grow cold was a shameful, even punishable, offense.)

That's where the old adage, "Where there's smoke, there's fire" originated, because if it smoked, it could usually be coaxed into a flame by gently blowing onto it, increasing the flow of oxygen and making the tinder hot enough to ignite. Once tinder was flaming, its short-lived flare could be fed and grown.

Today there are numerous types of tinder available, from Duraflame® tinder sticks and logs that light with the touch of a match and burn for up to thirty minutes to powder-able tinder cubes and even pastes. Cotton balls saturated with petroleum jelly or motor oil are home-made favorites, or the Fire Wicks made from cotton laundry string and saturated with molten paraffin (described in numerous of my own books and magazine articles). Any of these fit the definition of tinder, lighting easily, and burning long and hot enough to ignite kindling placed onto their flames.

Kindling

The next step is kindling, because without this crucial intermediate stage, tinder flames would just burn out before lighting larger chunks of fuel.

Most homes that burn wood have a store of kindling, small, finely split pieces of fire-starting wood, usually housed in a box next to a woodbox near a house's back door. A few generations ago, these boxes (which are often racks, nowadays) held the nightly supply of wood or, sometimes, coal. It wasn't uncommon to be awakened in the night by the creak of a cast-iron door, as someone returning from a trip to the outhouse re-stoked the fire, occasionally re-starting it from just a few embers using small lengths of kindling.

Thus was born another maxim: When a building blazes especially fast or fire fighters are pushed back by a forest fire, they often describe conditions as being "dry as kindling." Kindling is created by splitting larger pieces of firewood into over-sized match sticks that light readily and burn hot, becoming essential hot coals quickly.

Laying a Fire

Successfully making fire in a stove, fireplace, or forest lies in first preparing or "laying" it.

Wherever you're making a fire, whether it's in a fireplace or in a shallow excavated fire pit on the forest floor, it's a good idea to first lay a platform of small kindling sticks, arranged side-by-side, as close together as you can get them. On the ground, this configuration keeps fire from contact with evaporated moisture in the earth, which turns to steam and inhibits the growth of the flames. In a stove or fireplace, a platform also ensures good air flow under a fledgling fire because, no matter how compactly kindling sticks are pushed together, they're far from air tight and a fire growing atop them uses the platform for fuel.

Before you strike a spark or flame to the tinder, open the damper—the adjustable exhaust vent in your chimney—to its fully open position. If you're laying fire in a stove, open its draft—the adjustable air-intake vent that's normally located at the bottom of a firebox—completely. Fireplaces typically have no draft. With both these vents wide open, maximum oxygen reaches a growing fire, while the smoke it exhausts meets minimal resistance.

Lay your tinder atop the platform and set it to blazing. Lay a small (no more than half an inch across in on any plane) tinder stick directly across the flaming tinder. Lean another similar-sized stick across the first at a 90-degree angle. Next, lean another tinder stick across from the opposite direction, and so on. As you add tinder pieces from different directions, you'll note that a teepee (inverted V) formation emerging. As the first pieces begin to combust, keep adding more and progressively larger, sticks until a bed of red coals forms under the tinder.

When a bed of coals forms, switch from teepee fashion to parallel. One of the marks of an expert woodsman is that he only starts his campfire with the teepee method; once a good bed layer of red coals is created, he switches to the parallel form, or "Furnace Pile."

As a furnace pile throws more heat, it enables a greater volume of wood to be added to occupy the same space, combusts more efficiently and, most importantly, if you're living in a drafty old farmhouse, it burns longer. A furnace pile in a square-ish woodstove (as opposed to the rounded pot-belly style of old) enables the firebox to be stacked right to its top with fuel.

In most stoves, with the draft and the flue set to wide open, you'll hear the fire inside begin to roar as soon as the door is closed and latched. When you hear wood inside begin to crackle and pop, adjust its draft to less than half-open and do the same with the damper. Every stove is unique depending on many factors from stove size, type of fuel, prevailing winds, chimney diameter, and stove design. These generic adjustment instructions for draft and damper work with all of them, but experiment a little with your setup until you can set it to the optimum, with maximum heat from the coals, with minimum draft feeding them, and the correct amount of exhaust without either smoking out the room or allowing too much heat to escape as smoke.

Banking a Fire

There are many times, usually on a reallfy cold night, when you want a warming fire to last at least until mid-morning. It doesn't have to be hot, but it has to be smoldering for a long time, ready to be easily coaxed into a fire. A properly banked fire, whether it's in the woods or in an airtight woodstove, is equivalent to setting an engine to idle; not actively running at the capacity that it's capable of, but ready to return, with minimal coaxing, to its full potential in the shortest time possible. In a forest setting, a banked fire even survives a deluge, smoldering from its protected underside, able to be stoked back to a blaze in minutes, even with damp kindling sticks.

There are two critical components to banking a fire: a bed of hot coals and a section of wood, usually a whole, un-split log. Laid upon the coals, the log is too large to burst into flames and subsequently be consumed, but its underside smolders with hot embers until a gap caused by slow combustion, increases between the upper side of the fuel and the coals at ground level.

Chimney Care

It wasn't so long ago that probably most houses didn't have a central heating system, not even in the north, where there might be snow on the ground six months out of a year and ambient winter temperatures can be colder than 40 below zero.

Worse, there weren't even the "airtight" *Jotul*-type woodstoves; those didn't debut until the late '70s. Houses—sometimes big, two-story farmhouses, with single-pane windows and so many drafts that a good windstorm could blow out a candle in the living room—were heated by pot-belly, cast-iron heaters, often in conjunction with a fireplace. Those babies would glow red-hot on a cold night, eating split cordwood like a starving creature. But by morning they'd be cold as a winter night, with

A complete chimney brush kit; a necessity for anyone who burns wood.

not an ember surviving. There was a lot of family togetherness in those days, with the first person to rise re-stoking a new fire and everyone else gathered around, wrapped in blankets, sometimes wearing coats, boots, and gloves while the house heated up to a livable temperature again.

A square-shaped nylon chimney brush.

We hated those cast-iron monsters. From the time I was big enough to swing an axe, it fell to me to fill the woodbox with a night's supply of wood after school each day. On a cold night, those stoves might consume a quarter of a cord and we knew that, to be safe, there'd better be forty cords in the woodpile at the start of each winter. If we ran out—and that did happen a couple of times—it became a family chore to haul an old car hood into the woods on snowshoes and chainsaw enough fuel to last the next few days (don't kid yourself, no snowmobile has ever been capable of breaking trail through 4 feet of hard-packed snow).

The airtight woodstoves, with precision-controlled drafts and more efficient firebox designs, were a more than welcome innovation. Our firewood consumption literally dropped by 50 percent and when we returned home after a day's work, enough red coals remained in the firebox bed to ignite a new blaze. Their double-walled designs kept the stoves from becoming red-hot on the outside—something toddlers who still needed to learn the meaning of "hot, don't touch" appreciated, even if they didn't realize it. And adults appreciated the increased margin of safety provided by a burner that generated heated more efficiently without casting a dangerous amount of it against a wall.

Dangers of Creosote

But many of the same dangers remained. Burning green wood, especially pine, created a tar-like resin called "creosote." Creosote doesn't burn cleanly, but builds up on the walls of chimney flues and stove pipes until it becomes a thick layer that dries from continuous exposure to super-heated gases passing through the smokestack.

Dried creosote is dangerously flammable, even mildly explosive. Chimney fires are a genuine danger that once scared strong men the way children are frightened of what's under their beds at night. To wake up in the wee hours of a frigid night with an uncontrollable fire raging on the inner walls of your stovepipe and chimney, spewing chunks of burning tar from its top onto your roof, was terrifying. There was nothing for it, except to let it burn itself out and hope that it didn't burn through the thin metal walls of your stove pipe or set your roof ablaze before it did. Opening a stove's door caused it to burn out more quickly, but at the expense of making a chimney fire even hotter and a few people burned down their own houses—and their own bodies—with fiery explosions created by attempting to throw water onto the blaze.

Triple-wall stove pipe offers added protection from internal creosote-caused fires, because there are three layers of insulation, instead of a single wall of thin steel. Speaking personally, though, triple-pipe makes me a little nervous because it has been known to burn out entirely from the inside, while exhibiting little evidence of that from the outside. Inspect triple-pipe regularly, more frequently as time goes on, looking for spots of darkened, heat-discolored metal on its outer surfaces that would indicate its internal walls have burned-through. Replace the pipe at the first sign that its outer layer has been hole, regardless of how small the perforations might be.

Cleaning a Chimney

Because creosote does build up in a chimney over a period of time—faster if you burn unseasoned wood (that is, green wood that hasn't set in the elements long enough to thoroughly dry inside—a year is perfect), you'll need to clean your chimney. It was a rule of thumb in the old days that it be done once a year, at the start of the cold season, before using a stove or fireplace to which it was attached.

Salt and Newspaper

This is an old and simple method of cleaning creosote layers out of a chimney. Taught to me by an older lady from "Jaw-juh" when I was a young man, it made a believer out of my naturally skeptical self the first time I tried it. Simply pour about a tablespoonful of ordinary granulated salt in the center of two sheets of newspaper and then wad the pages up around the salt. Make about three of these wadded-up salt-and-newsprint packages, then toss them into the stove or fireplace being cleaned, and set them afire. As the pages burn with a green-blue flame, creosote falls from the chimney walls in chunks.

Chaining

Chaining is another simple technique for cleaning your flue that can be done by anyone capable of climbing onto a roof. With this method, a long—(about 16 feet of Grade 70 (5/16")) logging chain (available for about $30 at hardware and lumber outlets) is lowered down through the top of the chimney. There need not be (and should not be) any hooks or other attachments on the end, just a simple, plain chain. The chain is then rotated from above with a gentle swirling motion—just enough to cause its steel links to lightly scrape against the chimney walls, dislodging built-up soot and creosote. Be careful not to get too energetic about swirling the chain, because most chimneys are lined with clay tiles that can be cracked or broken with hard impacts.

Brushing

Brushing is a time-honored, but modernized, method of "sweeping" chimneys clean of creosote. The sooty profession was officially declared to be "lucky" after King William of Britain was pushed out the way of a runaway carriage by a lowly chimney sweep in 1066 (or thereabouts). King William was so grateful that he invited the sweep to attend his daughter's wedding. After that, the king announced that chimney sweeps should be afforded the right to wear top hats, a socially prestigious right previously reserved for wealth and nobility. The occupation was perhaps most immortalized by Julie Andrews in the song "Chim Chim Cher-ee," from the 1963 Disney musical *Mary Poppins*.

Chimney brushes retail for around $80, and resemble nothing more than over-sized rifle-bore cleaning rods. A metal- or nylon-bristle brush screws onto the threaded end of a long fiberglass or aluminum rod, to which additional threaded sections can be added if more length is needed.

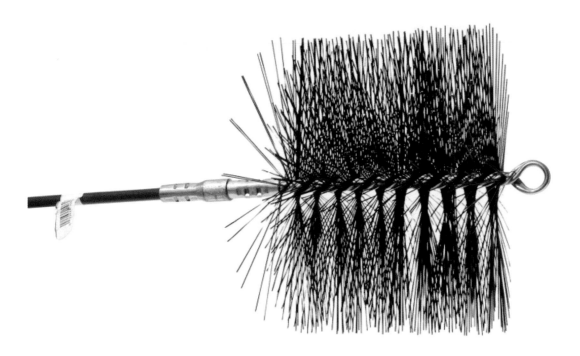

Available in a variety of sizes and configurations, a brush (or set of brushes) for cleaning soot and creosote accumulations from chimney flues is necessary.

Then the brush is simply pushed up and down the length of the chimney flue until the smokestack is clean.

Unless a flue is very dirty and neglected, a softer nylon-bristle brush is recommended over the wire version, as it's gentler on the bore. A wire loop at the opposite end of a typical brush enables a rope to be tied to it, then dropped down a chimney, where it can then be used to pull the attached brush after it. A heavy object—a short length of logging chain, apiece of railroad iron, a small rowboat anchor, even a barbell's weight disc—tied to the opposite end helps to gravity-feed through twists and bends; but use as little weight as necessary, and be gentle; don't let it slam against the flue walls. Alternately, a dog-sized chain can be clipped to the loop.

Roadblocks

It was a recreational fishing trip. No clipboards, no casting tracks, no photography; just my favorite open-face casting reel on a short custom pole, a couple of Mepps® spinner-type fishing lures, and my best fishing hole. Virtually no one knew this beaver pond was here, at the end of a rarely used telephone line access track. Fewer still owned the high-clearance 4x4 that it took to get back to where French Farm Lake terminated into French Farm Creek, from there into Lake Michigan. The creek was loaded with brook trout, but I was after the trophy-size largemouth bass that were simply common in the secluded beaver pond.

Clearing a roadblock; sometimes it's necessary to operate a chainsaw in awkward positions - such as cutting away one timber before you can reach another - be sure that you have a firm grip on your saw at all times, and try to see problems before they occur.

Cutting through 8 miles of fallen timber (about one tree every 50 yards) to reach people stranded at their cabin after a storm.

Sometimes you need to cut fallen timber that's lying directly on the ground,

but without getting damaging sand into your chainsaw.

Roadblocks know no season.

It was my second day there when I noticed a few dirty gray cumulous clouds gathering over the silhouette of the distant Mackinac Bridge. The Straits of Mackinac, connecting lakes Huron and Michigan, are infamous for generating their own, sometimes ferocious, weather systems: just like the Gulf of Mexico, only colder. There weren't tornadoes, as such, because dense forests along either shore prohibited them from forming, but waterspouts could wind up on the open water, and then slam into the shorelines with force enough to make how you defined them a moot point.

The cool of the night brought with it howling winds and hammer-force rain. I parked my truck as far out of reach of likely falling trees as I could and drove a half-dozen extra stakes around the perimeter of my tent. Even at that, the fabric domicile flapped like a flag in hurricane, and I heard several falling tree limbs bounce noisily against my truck.

Fallen trees can not only trap you out of the woods, but in them, as well—it pays to have the means to move fallen trees.

I broke camp the next morning, but got no more than a hundred yards down the trail before I encountered the first roadblock. It was good that I had elected to leave early, as nearly a dozen trees lay across my only route out of the forest, including one century-old behemoth that was three feet in diameter that had been ripped from the ground by its roots.

That incident illustrates one of the worst dilemmas you might encounter while truck camping or just driving through a forest. Little thought is given to being stuck *in* the woods by a fallen tree, but it isn't an uncommon occurrence, and even a relatively small tree can trap the most capable 4-wheel drive.

Roadblock Clearing Outfit

Because having one or many trees come down from lightning strikes, wind, or even mudslides is as likely as a flat tire on highway, it's just foolish not to guard against it any time you're going to be driving through the woods. A flat tire or engine trouble isn't the only trouble that can bring your vehicle to a standstill.

The Invaluable Axe

The most fundamental and important road-clearing tool is one that'll fit in the trunk of the most compact cars: Once again the lowly axe appears here in a starring role. Axes have been clearing the

An axe is the quintessential roadblock buster, and there aren't many objects that won't yield to this oldest of lumberjacking tools.

way for humans since before smelted metals existed, and with today's advanced alloys and super-strong handle materials, it's more efficient than ever. There are few blockades that cannot be broken with an axe.

But as axes have improved over time, though the number of people who are proficient with them have dwindled. The concept behind chopping wood is to remove as much of the material as possible with the fewest blows and the least expenditure of energy—not many chores are more bone-wearying than chopping wood.

When halving a tree trunk that's more horizontal than vertical with an axe (chopping down a standing tree needs a slightly different technique), begin with an inward-angled cut from either side. The space between each initial cut should be equivalent to three-quarters of its diameter. After the second cut, give the axe handle a violent twist downward to break free a large chip of wood between the two angled cuts.

Another pair of angled cuts driven just inside the first two, followed by another strong twist of the axe's handle breaks free another large chunk of wood. Two more inward-angling cuts, another twist, and so on until the log is parted.

You'll note that the cut naturally becomes more V-shaped the deeper it penetrates—that's the reason for starting with such a wide cut.

Using a Chainsaw to Remove Roadblocks

Immortalized by Gordon Lightfoot in his song "The Wreck of the Edmund Fitzgerald," after that gigantic ship went down with all twenty-nine hands in 1975, the Gales of November are a very real phenomenon to people living along the shore of Lake Superior. The storms were perhaps a little too real to the couple that we were even now driving out to rescue from their remote cabin—8 miles from the nearest power line, a half mile from the nearest dirt road.

They'd moved here the previous spring, visions of that old movie *The Wilderness Family* dancing in their leather-hatted heads. They'd moved to Arizona before the end of their first winter, and the fact that we were now cutting our way to them, with a downed tree every 50 yards or so, was one of the bigger reasons for the short duration of their stay.

Although the machine is a champion of roadblock removal, having a chainsaw means packing along an entire kit for it, including blended gasoline, bar oil, wrenches, screwdriver, and a chain sharpener. Without even one of these items, a conventional gasoline-powered chainsaw might soon be useless; even the new battery-operated models—some of which our field trials have shown to be ideal for such emergency purposes—need lubrication oil for the chain.

Other common-sense safety items might include eye protection, heavy gloves, Kevlar™ chaps, and a hardhat. If you've included a chainsaw, you should consider these safety items to be a necessary part of that system.

When operating a chainsaw, it's impossible to be too cautious. According to the Consumer Product Safety Commission, legs are up to five times more likely to be injured in chainsaw mishaps than any other parts of the body. Be especially aware of kickback, that notorious tendency for a spinning chain to catch and cause the saw to suddenly spring back toward its user.

Dealing with the infamous Gales of November, Cutting our way out 8 miles to rescue a couple stranded in their cabin.

Ironically, in the remote areas that need them most often, there are usually few rescue resources, you can't count on Big Brother riding to your rescue in an emergency.

Storms can come in any weather.

Kickback can happen at any time, so that's when it should be expected. Never operate a saw with one hand and always stiff-arm the handle so that even a strong kickback can only cause the machine to recoil over your head. There are times, especially when it's necessary to trim branches from a tree trunk before you cut it in half, that you must cut with the nose of a chainsaw; kickback is almost a given under these circumstances, so be prepared.

These snow-bent saplings show why trees in snow country are prone to growing with twisted grain that makes then hard to split for firewood.

Note the "springback" sapling being held down by the weight of this wind-felled white pine; springbacks have killed or injured many a lumberjack.

The tree went right where it was intended to go, but beware the springback sapling to the right,

being held down under tension by its weight.

Also be aware of what Husqvarna calls a "springback." A springback occurs when (usually) wind blows down a large tree—a common sight in wet forests, where sometimes huge trees are uprooted and just topple over—and they land on springy green saplings that bend to the ground

Whether you live in timber country or a suburb, felled trees can block your path.

under their massive weight.

As the moniker implies, restrained saplings may suddenly spring back when weight on them is removed, or even reduced, knocking people out of their way with violent, even fatal force. Look for springbacks before you cut any fallen tree, and always cut those first. And, of course, be aware that they may be suspending several hundred pounds of wood, which might suddenly fall if a springback is severed.

Extending the winch cable to its fullest length.

Winches and Ropes

The same hand winch that's so invaluable for felling trees can be at least as indispensable for breaking a roadblock. Again, it should be rated for at least 2 tons and mated with a good timber-rated rope, maybe a tow strap. All lines should be in good condition, with no fraying or obvious damage that might weaken them.

When it becomes necessary to move a downed log with just muscle, you'll note that getting a grip so that you can actually use your physical strength to best advantage is very difficult.

A log that you can't lift using your hands can often be lifted by first rolling it onto a rope (the easiest way to get a rope wrapped around its girth), then tying it off with a combination slipknot and half-hitch—the "choker hitch" described previously. Then, by tying the free end of the rope to a stout section of tree limb, an axe, or shovel handle, you create a way to lift perhaps double the weight that you could using just your hands. By lifting just one end of a log then pulling backward, you can move an impressive amount of weight with minimum exertion.

FIRST AID

The young man from Boyne City, Michigan, actually knew better; he'd grown up around lumber-jacks and loggers and knew the folly of pruning overhead branches with a chainsaw. But like many young bucks in a group of their fellows, he was overcome by the drive to impress his companions with a prowess that was not supported by sufficient experience. He lacked the foresight to avoid the kickback that occurred when his saw caught against a knot, and snapped downward to cut deeply into the muscle of his left shoulder.

How close that young man came to death is a matter for conjecture, but it's a sure thing that being driven out of the woods in a bouncing 4x4 pickup, then racing to the nearest hospital 20 miles away while he held a greasy rag over his own hemorrhaging wound wasn't the optimum in first-aid treatment.

That wasn't the first time a serious accident had happened to someone deep in a forest. Sometimes it has been a slip while gutting a deer, sometimes when an axe skipped off a hard knot, or any of the virtually innumerable injuries that can and too often do happen to victims who didn't see it coming. When serious trauma occurs in the woods, where cell-phone coverage can be iffy to nonexistent and directing medical personal to a location by phone can be a challenge—even with GPS—everyone concerned can benefit from having an all-business emergency medical kit.

For help with this project, I turned to twenty-four-year veteran paramedic Cheanne Chellis, who has lost count of the patients she's helped to keep breathing while bouncing along in the back of an ambulance along rutted two-track roads in Michigan's wild Upper Peninsula. The "jump kit" medical bag she carries in her own truck while driving the wilderness roads around her home near Paradise, Michigan, is a good guideline to assembling a backcountry trauma outfit that might in fact save someone's life.

Quik-Clot Sport Anti-Hemorrhagic Bandages

Easily one of the most lifesaving medical innovations of the past decade are the anti-hemorrhagic blood-clotting agents, which can staunch blood loss from a severed artery without need of a tourniquet that can kill an entire limb, necessitating amputation. First used in combat and on animals (including our own sled dogs), products like Z-Medica's Quik-Clot were available only to medical professionals until 2010, when the civilian-approved sport line of pre-treated Quik-Clot "four-by-four" (3.5x3.5 inches, actual) bandages became available over the counter. Removed from their vacuum-sealed foil envelopes and taped directly over a severed artery, gunshot wound, or any profusely bleeding injury, either Quik-Clot Sport or the new antiseptic Sport Silver has a battle-proven ability to save life and limb when a wound is bad and help is far away. Available in 25- or 50-gram sizes, prices for Quik-Clot Sport 25 average about $10 per envelope, around $12 for Sport Silver.

Paramedic-Approved First-Aid Kit

Being human means being fragile. It's reassuring to think of ourselves as rugged, but the truth is that it's pretty easy to kill a human being. An injury that ruptures or severs an artery can bleed us to death in only four minutes. A hard knock to the brainpan can cause fatal cerebral hemorrhaging. There are more ways to kill a human than there are to flay the proverbial cat.

The odds of surviving a potentially fatal trauma have risen dramatically in recent years. Advances in medical science have given even laymen like myself a better understanding of human physiology and spawned a host of potentially lifesaving tools that couldn't have existed for our predecessors even a generation ago.

I ponder such things when I'm alone in the woods and medical help is a half-day away. But even in a city, with an ambulance available in just minutes, prompt, proper application of up-to-date medical tools can save lives. To address that concern, I turned to the paramedics of Allied Ambulance, whose daily rescues might encompass a multi-car pile-up on Interstate 75 to wilderness extractions from the nearly two million acres of surrounding public forests.

An essence of professional experiences and lessons learned (mostly vicariously for me, thankfully), here's the first-aid kit that you'll find in my backpack today:

1 bottle ibuprofen, 50 count

1 4x4 gauze sponge

1 large roll gauze

1 roll 1"-wide self-adhering tape

1 elastic wrap

1 sterile suture kit

1 chewable Pepcid AC heartburn tablet

2 alcohol prep pads

2 Quik Clot™ anti-hemorrhage sponges

6 sealed diphenhydramine Hci (Benadryl) capsules

6 loperamide Hci anti-diarrheal capsules

1 triple-antibiotic ointment, tube or envelope

1 miniature LED headlamp

1 penlight

1 toenail clipper

1 tweezer

1 bandage scissor

1 small magnifying glass

1 first aid manual, Emergency First Aid (American Red Cross)

GLOSSARY OF LUMBERJACKING TERMINOLOGY

Axe: Any wood-chopping tool; axe sizes, grinds and edges differ depending upon preference and the chopping job.

Banana Grind: An axe blade ground thinner on the edges and thicker in the middle, resembling a banana.

Barber Chair: A tree which splits upward along the grain during falling.

Block: An un-split section of firewood.

Bone: Hard or "tight" wood.

Boom: A corral to hold free-floating logs, chained together in the water until ready to move downstream.

Boom Run: A head-to-head competition featuring two opponents racing across floating, linked logs. The event's origins come from a need for lumberjacks to corral timber in ponds by running across the booms.

Branding Ax: A tool used for marking ownership of a log.

Buck: To cut a tree into lengths after it has been felled.

Bucker: One who saws trees into logs.

Bullcook: (Also known as the *crumb boss*), a boy who performs chores around camp.

Calks or **Caulks**: (Pronounced "corks" by rivermen), sharp, short spikes set in the soles of boots to prevent the men from falling off logs. Also sometimes used on horseshoes.

Cant: Any rounded or squared log.

Cant Hook: A wooden lever with a hooked arm near its lower end which passes over the log, grips it, and affords a grip by which it can be turned over. Also called a cant dog.

Chainmail Sock: Protective socks worn by professional lumberjacks during the chopping events to keep them safe from injury.

Chaps: During chain saw events, competitors wear these covers featuring cut-retardant material to protect legs.

Chisel Grind: An axe blade with a steep, uniform grind; used for working axes.

Conks: Fruiting bodies of fungus in rotting wood.

Cookee: A lumbercamp cook's assistant.

Cookie: The circular wooden discs cut during sawing events.

Corks: Short, sharp spikes set in the soles of shoes.

Crotch Line: A device for loading logs onto railroad cars.

Crown Fire: A forest fire that reaches into the tops of trees.

Cutout: Disqualification in a sawing event when a full disc isn't cut, or a competitor cuts over, or on, the allotted guideline.

Deacon Seat: A bench, made from a large log split lengthwise, running the length of a bunkhouse

Donkey: A stationary multiple drum machine, powered by steam until the prevalence of the internal combustion engine.

Drag Day: The point in the work month when a man can get an advance on his wages.

Driving Blows: Long, powerful axe swings that finish the initial side of a chopping block before turning to sever a section of log entirely.

Driving Pitch: High water suitable from driving logs down a river.

Duplex: A stationary engine that both assembles (*yards*) and loads logs.

Flat Grind: An axe blade ground flat from corner to corner. Generally not seen on working axes, only competition axes, because it is too fragile.

Foot Block: Wooden foot braces nailed into the deck to provide a lumberjack with foot traction during the single buck event.

Gandy Dancer: A pick-and-shovel man.

Gin Pole: A short spar, used for loading and unloading logs.

Gyppo: Contract worker who works by increments—feet, loads, miles, etc.

Hardtack Outfit: A company running a logging camp which provides substandard food (derived from the cheap and long-lasting cracker or bread of the same name).

Hayburner: A horse.

Helicopter Turn: A competitive lumberjack chopping technique popularized by Dave Jewett where an underhand chop competitor jumps, with axe raised above, and turns to begin backside cut in one fluid motion.

Heel: The edge of the axe that is nearest to the handle or bottom of the axe blade.

Highball: To hurry.

Hollow: The area of relief, or depression, in the axe behind the chisel or banana that provides lift to bust chips out of the way as the axe travels into a block.

Hoot-Nanny: A small device used to hold a crosscut saw while sawing a log from the bottom up.

Hovel: A stable for logging horses or oxen and also a group of camp buildings.

Ink Slinger: A logging camp timekeeper.

Into the Small Wood: When a chopper has cut to the center of a block, smaller and tighter chips are produced due to the angle of axe and density of the interior wood.

Iron Burner: The camp blacksmith.

Jerk Wire: A line attached to the whistle on a yarding donkey, by which a young man (a *punk*) blows starting and stopping signals.

King Snipe: The boss of a track-laying crew.

Landing: The place where logs were gathered for loading or to be later rolled into the river.

Long Logger: A logger working in the fir and redwood country of the Western U.S., where logs were often cut in lengths up to 40 feet.

Macaroni: Coarse wood shavings.

Melon of a Block: Very soft wood; often used in conjunction with the phrase "so soft you could see seeds coming out."

Misery Whip: A single buck-sawing event.

Mulligan Car: A railroad car where lunch is served.

Nosebag: A lunch bucket.

Nosebag Show: A camp where the midday meal is taken to the woods in lunch buckets.

Peavey Hook: A type of cant hook having a spike at the end of the lever (also known as *cant dog*); a tool with a sharp point and a movable hook on it, used on a river to create leverage when moving floating logs.

Pecker Pole: A small tree, often found in the understory of old growth.

Peg and Rakers: The cutting and pulling teeth found on a cross-cut saw.

Pickaroon: Used in pulling small timbers out of the water; broken axes were often made into pickaroons.

Pike-Pole: A long pole with a sharp spike on the end used to control logs floating on the water.

Potlatch: A social gathering (said to be a Chinook term).

Powder: Dynamite

Pulaski: A thick-handled tool with oval eye used as a combination axe and hoe, named after its inventor.

Pulp-Hook: A short, curved, steel hook used with one hand to draw a stick of pulpwood.

Road Monkey: The man with the job of keeping the sled road clean.

Sagging Board: A springboard that is not level or tilted slightly upward, often a result of a bad pocket.

Scaler: The person who estimated the number of board feet in a log.

Schoolmarm: A log or tree that is forked, stable in river driving because it does not roll easily.

Scoop Hit: When the side of the axe hit the wood instead of the blade or edge.

Seconder: The second person aiding a single buck participant by wedging or oiling the saw blade.

Shin and Foot Guards: The aluminum guards worn by collegiate competitors to protect feet and shins during the underhand chop.

Short Cut the Front: The practice of mistakenly taking too few chips out of the front face of a block.

Short Staker: (Or *boomer*): a worker who quits after earning a small sum.

Short Stroking: Failing to pull the entire length of peg and rakers through a block of wood.

Skidroad: The path over which oxen pulled logs; it came to mean the part of a city where loggers congregate.

Slab: Uncut wood.

Slab Rule: The required number of nails inserted into the wood to keep the wood in place with the opening blows.

Slabbing Nails: Nails used to keep uncut wood in place.

Slash: Tree tops left on the ground after a logging operation; often the cause of "slash fires."

Snoose: Damp snuff or chewing tobacco.

Snubber: A device for braking sleighs as they descend steep hills.

Sougan: A heavy woolen blanket.

Spar: The wooden pole climbed in the speed climb event.

Stick of Wood: Prepared competition wood.

Swedish Fiddle: A crosscut saw.

Three-Cutters: A single buck saw featuring three cutting teeth or pegs, and one raker.

Throwing the Chain: A potentially dangerous event most common in a hot chainsaw when the chain comes off the bar or breaks altogether.

Tight Wood: Logs producing few chips usually found at the compressed, lower part of the tree.

Timber!: Warning that a tree is about to fall.

Tin Pants: Waterproof clothing worn by loggers in the rainy Pacific Northwest.

Toe: The edge of the axe that is farthest from the handle or top of the axe blade.

Turkey: Any kind of a sack in which a lumberman of old carried his belongings.

Twitch or to **Snake a Log:** To have a horse drag (twitch) a log along the ground without having to load it.

Two-Cutters: A single buck saw featuring two cutting teeth or pegs, and one raker.

Wedger or **Oiler:** (See seconder), the person who tends to a single buck competitor's saw during a competition.

Widowmaker: A branch or tree that is held suspended above ground, often a cut tree that has hung-up in the branches of another tree, ready to fall onto the unwary.

MANUFACTURER RESOURCES

Chaps

USDA FS, Missoula Technology and Development Center
5785 Hwy. 10 West
Missoula, MT 59808–9361
Phone: 406-329-3978
E-mail: wo_mtdc_pubs@fs.fed.us
Website: www.fs.fed.us/t-d

CitroSqueeze
CDR Chemical, Inc.
16182 Gothard St., Suite J
Huntington Beach, CA 92647
Phone: 888-270-4237
Website: www.cdrchemical.com

Solutions Safety Services, Inc.
1516 E. Edinger Ave. Unit A
Santa Ana, CA 92705
Phone: 714-849-5653
Fax: 714-843-6743
Website: www.solutionssafety.com

Seam Grip
McNett Corporation
1411 Meador Ave.
Bellingham, WA 98229
Phone: 360-671-2227
Fax: 360-671-4521
Website: www.mcnett.com

Chainsaw Manufacturers

Craftsman: Sears Brands LLC. www.craftsman.com

Dolmar: Part of the Makita group (formerly known as Sachs Dolmar). www.dolmar.com

Echo: Part of Yamabiko Corporation. www.echo-usa.com

Efco: Part of the Emak Group. www.efcopower.com

Homelite: Part of the TTI group. www.homelite.com

Husqvarna: Top-level brand of the Husqvarna Group (Husqvarna AB). www.husqvarna.com

ICS (concrete cutting chainsaws): Division of Blount Incorporated. www.icsdiamondtools.com

John Deere: www.deere.com

Jonsered: Part of the Husqvarna Group. www.jonsered.com

Makita: Top brand of the Makita Group. www.makita.com

McCulloch: Part of the Husqvarna Group. www.mccullochpower.com

Oleo-Mac: Part of the Emak Group. www.myoleo-mac.com

Olympyk: Part of the Emak Group (see Oleo-Mac above).

Partner: Part of the Husqvarna Group. www.partner.biz/int

Pioneer: (no longer available) Part of the Husqvarna Group.

Poulan: Part of the Husqvarna Group. www.poulan.com or www.poulanpro.com

RedMax: Part of the Husqvarna Group. www.redmax.com

Remington: Currently makes electric models only. www.remingtonpowertools.com

Shindaiwa: Part of Yamabiko Corporation. www.shindaiwa.com

Solo: www.solo-germany.com or www.solousa.com

Stihl: Top brand of the Stihl Group. www.stihl.com

Victus: Part of the Emak Group. www.emakgroup.com

INDEX